MECHANICS in A

Modelling and Practical Investigations

Mike Savage

Senior Lecturer in Applied Mathematics
University of Leeds

Julian Williams

Lecturer in Education
University of Manchester

Cambridge University Press

Cambridge
New York   Port Chester   Melbourne   Sydney

Published by the Press Syndicate of the University of Cambridge
The Pitt Building, Trumpington Street, Cambridge CB2 1RP
40 West 20th Street, New York, NY 10011, USA
10 Stamford Road, Oakleigh, Melbourne 3166 Australia

© Cambridge University Press 1990

First published 1990

Printed in Great Britain by Bell and Bain Ltd., Glasgow

British Library cataloguing in publication data
Savage, Mike
Mechanics in Action.
1. Mechanics
I. Title   II. Williams, Julian
531

ISBN 0 521 38941 0

LRT registered user Number 90/1138

Map on page 21 reproduced by permission of Michelin from
their Motoring Atlas of Great Britain and Ireland.

MU

CONTENTS

ACKNOWLEDGEMENTS

We wish to thank the many teachers who, on in-service courses and in working groups associated with the Mechanics in Action Project, have trialled various materials, ideas and activities. Particular thanks go to Ann Kitchen, Carol Martin, Liz Nash, Sandra Hoath, Val Aspin, Lucy Lammas, Veronica Johns, Phil Osborne and Eric Aldred without whose support and feedback the work would have been impossible.

A number of other collaborators on INSET courses have provided ideas and suggestions, namely Tom Roper, Janet Jagger, Phil Gaskell and Tim David.

Special thanks are due to Les Goulding, who laid out the worksheets for trialling in schools in 1988.

INTRODUCTION

Mechanics has always been fundamental to the study of mathematics, science and engineering and is as relevant today as at any time since Newton. There *is* a need, however, to regenerate and inspire the teaching and learning of the subject in schools and colleges, where its presentation is often dull, abstract and didactic.

At this time there is also a need to develop A-level as a whole, in the light of recent changes at GCSE level. The aim of the Mechanics in Action Project and of this book is to bring the practical and investigative spirit of GCSE to bear on real problem solving in Mechanics. At A-level, mechanics is, in our opinion, an ideal subject to study for the following reasons:

- mechanics provides ample scope for real problem solving and for teaching modelling skills through practical investigations.
- mechanics can motivate the pure mathematics of functions, calculus and geometry.
- mechanics forms a natural link with science and technology.

This book is intended for teachers who have yet to experience in their classrooms the richness of practical investigations in mechanics. It provides a set of teaching resources, in part B, and a rationale for and background to the modelling approach, outlined in part A.

WHY MODELLING? WHY PRACTICAL INVESTIGATIONS?

The simple answer is that teaching students to solve real problems with mechanics involves both! Modelling is essential to the problem solving process; practical investigations provide the ideal vehicle for **teaching** students to model. Real problem solving involves a shift in emphasis from stage 2 towards stages 1 and 3 of the modelling process:

Much more time needs to be spent on identifying challenging problems, building and refining appropriate models and interpreting and validating solutions. This will, in our opinion, have to be done at the expense of practising standard solutions to standard (though technically demanding) examination questions.

Our approach to real problem solving is to provide students with practical investigations using simple apparatus. This encourages students to be actively engaged from the beginning in the formulation of interesting problems. This, in turn, provides strong motivation for subsequent investigation, analysis and solution. The apparatus and worksheets help students to identify important features so as to make assumptions and introduce variables, in other words, **to set up a model**. Having applied Newton's laws to a particular problem and obtained a mathematical solution, students are then encouraged to **interpret and validate** their solutions, using the apparatus.

EMPIRICAL MODELLING AND NEWTONIAN MODELLING

Mechanics is an ideal vehicle for teaching the principles of modelling and enabling students to acquire modelling skills. There are two main reasons for this. Central to mechanics are Newton's laws, a sound body of scientific knowledge, which gives mechanics a coherence often lacking in some other fields of enquiry. Secondly there is an enormous variety of applications generating interesting problems and interesting features to investigate.

Many physical situations can be **modelled empirically** without understanding Newtonian theory, simply by collecting data on relationships between quantities and modelling these with functions and calculus – **pure mathematics**. See, for instance, 'Distance, speed, acceleration' (chapter 8), 'Timing oscillations' and 'Simple pendulum' (chapter 10), and 'Bouncing ball (2)' (chapter 11).

In fact, the stock models of mechanics which provide laws of force for tension and friction make ideal empirical modelling investigations for students of pure mathematics and statistics. Examples of this kind include 'Three masses' (chapter 5), 'Balancing a ruler', 'Beam balance' (Chapter 6), 'The law of friction' (chapter 7), 'Hooke's law' (chapter 10) and 'Bouncing ball (1)' (chapter 11).

Whenever data is collected in an experimental situation there are **statistical investigations** of interest: 'How do you find a straight line or curve of best fit?', 'How do you decide if the time period of the oscillation is **significantly different** when the mass is changed?' 'What is the range of error expected in validating a prediction?' However, the power of modelling in mechanics only becomes fully realised when it is based on Newton's laws. Selecting appropriate force laws together with an appropriate geometry of the physical situation, one can build powerful predictive models which are rich in interpretation. There are many of these to be found in part B of this book.

Often the situation being investigated can be treated using either an **empirical model** or a **Newtonian model**. For instance, 'Timing the oscillations' (chapter 10) can involve modelling empirically with a formula: $T \propto \sqrt{m}$, or using the Newtonian model for simple harmonic motion which gives

$$T = 2\pi \sqrt{\frac{m}{k}}$$

A modelling investigation can involve identifying a number of problems associated with a situation and solving them using empirical and/or Newtonian models. It may begin with a real situation outside the school, such as at the fairground or in industry, and subsequently involve related practical work in the classroom; see chapter 9 for a number of such examples.

Judgements should always be formed as to the validity in practice of any model, and these will often lead to the need for a refinement of the model.

A GUIDE TO MECHANICS IN ACTION, MODELLING AND PRACTICAL INVESTIGATIONS

The development of modelling investigations at A-level need not make intolerable demands on the teacher. Many simple practicals are described in part B. It is more the **approach** to modelling which is important. If the teacher presents mechanics as a set of recipes or regards the theory as engraved on tablets of stone then the pupils will not learn how to model.

For this reason, we provide background material in part A which will help the teacher to see Newtonian mechanics afresh in terms of modelling ideas. Chapter 1 presents the scientific model of Newton's laws and shows how all the other principles of mechanics follow as consequences of this model.

Chapter 2 examines the stock models for gravity and other forces, together with the stock models that Newton used for the Earth. The validity of different models for pulleys, strings and springs is also examined. Without understanding the assumptions on which all these models are based, students cannot appreciate the nature of the models they will use when solving problems.

Chapter 3 gives examples of how students should use the 3-stage modelling process when solving real problems. All the solutions to the problems in part B are written up in this format.

Chapter 4 provides examples of common difficulties and misconceptions which students experience with the principles and models of A-level Newtonian mechanics. A set of diagnostic questions and strategies for overcoming the difficulties are described.

Part B contains teaching resources for practical investigations on force, moments, friction, kinematics and energy, circular motion, springs, oscillations and simple harmonic motion and impact. The practicals are in no particular order within the chapters, and can be used in various ways with different students on different courses. In the introduction to each chapter we have described some of the ways each have been used, and most practicals have notes on aims, equipment, plans, solutions, misconceptions and extensions.

There are some references mentioned in the text. These can be found in the Reference section at the end of the book.

Some practicals have been written up at greater length to give extra help and support. If you have never taught 'practical mechanics' you could start with these:

Bathroom scales and a brush
Balancing a ruler
The ruler problem
The dangerous sports club problem
Distance, speed, acceleration
Pennies on a turntable
Conical pendulum (1) and (2)
Timing the oscillations
Bouncing ball (1)

All these have proved particularly simple and successful with teachers and pupils, and are very rich in ideas and extensions. They mostly use apparatus easily available in school. However, Unilab produce kits of equipment which have been used with these materials and will save you the problem of finding class sets of bouncing balls, stopwatches, springs, etc. They also market the Leeds Mechanics Kit, which is referred to in the text and is ideal for many of the circular motion and simple harmonic motion practicals as well as the friction and parallelogram of forces experiments. Unilab can be contacted at Unilab Ltd, The Science Park, Hutton Street, Blackburn, BB1 3BT.

PART A
Background to modelling

If it universally appears, by experiments and astronomical observations, that all bodies about the Earth gravitate towards the Earth . . . in proportion to the quantity of matter that they severally contain; that the Moon likewise . . . gravitates towards the Earth . . . and all the planets one towards another; and the comets in like manner towards the Sun; we must, in consequence of this rule, universally allow that all bodies whatsoever are endowed with a principle of mutual gravitation.

Newton, *Principia*, 1687

To tell us that every species of things is endowed with an occult specific quality by which it acts and produces manifest effects, is to tell us nothing. But to derive two or three general principles of motion from phenomena, and afterwards to tell us how the properties and actions of all corporeal things follow from those manifest principles, would be a very great step in Philosophy, though the causes of those principles were not yet discovered.

Newton, *Opticks*, 1730

Fig. 1.1 Sir Isaac Newton

aims of chapter 1

After careful study of this chapter you should:

- appreciate the historical perspective in which the work of Johannes Kepler provided a challenge to Newton.

- appreciate why Newton proposed a law of gravitation and three laws of motion.

- understand the essential concepts of 'particle', 'equilibrium' and 'force' to which Newton's laws refer.

- understand how Newton validated his inverse square law of gravitation.

- understand how Kepler's observational laws can be derived from Newton's laws.

- appreciate the validity and the limitations of Newton's laws.

- understand the key concepts of momentum, impulse, energy, work and how they relate to force.

- know how the conservation of momentum for a particle or a system of particles or a rigid body **follows** from Newton's laws.

- know how Newton's second law for a system of particles or a rigid body follows from Newton's laws.

- know that the conservation of energy for a particle and a rigid body is derived from Newton's laws when any force doing work on the body is conservative, for example gravity.

- know that the total energy of a system of particles is **not** generally conserved.

1.1 NEWTON'S LAWS

Throughout recorded history astronomers have observed the movement of the planets against a background of the 'fixed stars' and sought to explain their observations. Likewise there have been many attempts to understand the motion of terrestrial bodies. As one theory superseded another the ancients consistently assumed that planetary and terrestrial motions were essentially different – goverened by different causes or mechanisms. The genius of Isaac Newton was his perception of how to unite in one general theory both the movement of the planets and that of terrestrial bodies.

1.1(a) The law of universal gravitation

Isaac Newton was educated at the King's School, Grantham until 1661 when he won a scholarship to Trinity College, Cambridge, to study the natural sciences. As an undergraduate he became aware of the work of Johannes Kepler who had published three 'Laws' for planetary motion based on observational data (see References, Kepler (1), (2)). Kepler asserted that:

1 the path of each planet P is an ellipse with the sun S at one focus
2 the line joining sun to planet sweeps out equal areas in equal intervals of time.

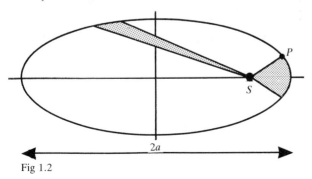

Fig 1.2

3 the square of each planet's period, T, is proportional to the cube of the semi-major axis of the ellipse, a,
$$T^2 \propto a^3.$$

An important problem for seventeenth century science, following the work of Kepler, was to identify the 'force' which maintained each planet in orbit about the sun. A magnetic force was suggested by William Gilbert (see Roller (3)), an idea which was also taken up by Kepler, but it was Isaac Newton who conceived of a force due to gravity varying inversely with the square of distance. In modern terms the law of gravitation may be stated as follows:

The law of universal gravitation

Every particle in the universe attracts every other particle along their line of centres.

If two particles of mass m_1 and m_2 are a distance d apart, then the force of attraction has magnitude F where

$$F = \frac{G m_1 m_2}{d^2}$$

and G is a universal constant

$$G = 6.67 \times 10^{-11} \mathrm{m^3 kg^{-1} s^{-2}}$$

The idea of a universal gravitational force emerged as a result of observations that:
(a) the Earth acts on both a falling apple, and on the moon;
(b) Jupiter and Saturn act on their moons;
(c) the sun acts on each of the planets in the solar system.

Newton gives the following account of his discovery of the law of universal gravitation.

> 'I began to think of gravity extending to ye orb of the moon and . . . from Kepler's rule [$T \propto a^{3/2}$] . . . I deduced that the forces which keep the Planets in the Orbs must vary reciprocally as the square of their distances from the centres about which they revolve: and thereby compared the force requisite to keep the moon in her Orb with the force of gravity at the surface of the Earth, and found them to answer pretty nearly.'

(See section 1.1(c).)

With hindsight we can see that the discovery of this inverse square law of gravitation was undoubtedly a crucial step in the development of Newtonian mechanics. On its own, however, it was insufficient to formally derive (and thus confirm) Kepler's observational laws, since there was no relation between force and motion. There was a need to determine the motion of a body when acted upon by an applied force. The problem is well illustrated by observing that the Earth acts on both a falling apple and the moon, yet their motions are quite different. It is perhaps not surprising therefore that twenty years elapsed before Newton published his three laws of motion in the *Principia* in 1687. (See References, Newton (4).)

1.1(b) The laws of motion

If we consider a particle, of mass m, whose velocity is **v** and acceleration **a** when acted upon by an applied force **F**, then Newton's three laws of motion may be stated in modern terms as follows:

Newton's first law : NL1

Every particle continues in its state of rest or of uniform motion in a straight line unless it is compelled to change that state by the action of an applied force.

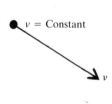

Newton's second law : NL2

The rate of change of linear momentum of a particle is directly proportional to the resultant applied force acting on the particle:

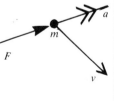

$$\mathbf{F} = \frac{d}{dt}(m\mathbf{v}) \qquad (1.1)$$

and if m is constant then

$$\mathbf{F} = m\frac{d\mathbf{v}}{dt} = m\mathbf{a} \qquad (1.2)$$

Newton's third law : NL3

Whenever two bodies interact they exert forces on each other which are equal in magnitude and opposite in direction. So whenever body A exerts a force **F** on body B, B exerts a force −**F** on A.

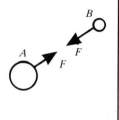

Newton's laws use the terms 'particle', 'equilibrium' and 'force', which require some explanation.

Bodies or particles?

It is noted that the first two laws refer to a particle, by which is meant a **point mass**, a concentration of matter which occupies a single point in space. There are good reasons for introducing 'particle' despite the fact that Newton used the word *corpus* (body) in the *Principia*.

- Many bodies in the real world can be adequately modelled by particles.
- The mathematical formulation of problems with particles is much simpler than that for bodies having size and shape.

The question of when it is valid to model a body by a particle is discussed in chapter 2.

Equilibrium

A body is said to be in equilibrium if no resultant force acts upon it.

A body (particle) is in equilibrium if it is at rest – 'static equilibrium', or if it moves with constant velocity – 'dynamic equilibrium'.

Two illustrations of dynamic equilibrium are:

1 A spacecraft with engines switched off moves in deep space where the gravitational attraction by other bodies (planets, comets, meteorites) is negligible. Since no force acts, the spacecraft moves with constant velocity, **v**.

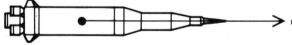

Fig. 1.3

2 An ice puck moves upon the ice and slows down, but only very slowly. If we now imagine the ice to become increasingly smooth, then in the limit, friction vanishes and the puck will move with constant speed in a straight line.

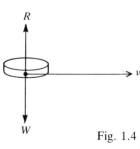

Fig. 1.4

In this case no resultant force acts yet the puck is subject to individual forces – weight **W** and reaction **R** – which are equal in magnitude and opposite in direction.

Force

What do Newton's laws tell us about force?

- The first law tells us that the effect of an applied force is to disturb equilibrium and change the velocity of a particle.
- The second law tells us that the consequence of an applied force is to change linear momentum and so produce an acceleration.
- The third law tells us that applied force arises from the **action of another body**.

The characteristics of force may be summarised as follows:

> **Force:**
> - is the action of one body upon another.
> - disturbs equilibrium and causes acceleration!

1.1(c) Validation of Newton's laws

In order to validate his laws of mechanics, Newton needed to account for Kepler's observational laws and to make predictions about the motion of other bodies which could be confirmed either by experiment or by known data. In this section we shall consider Newton's geometrical arguments for validating the inverse square law and then give a modern derivation of Kepler's laws.

Validation of the inverse square law

To prove that the same gravitational force caused both the fall of an apple and the acceleration of the moon towards the Earth, Newton had to:

- know the acceleration due to gravity of the apple at the Earth's surface. (Let its magnitude be g.)
- calculate the acceleration of the moon. (Let its magnitude be g_m.)
- find a relationship between these two accelerations and check that this was in agreement with an inverse square law of gravitation.

The acceleration at the Earth's surface was well known; experiments by Galileo and Huygens had established a value for g:

$$g = 9.8 \text{ m sec}^{-2}$$

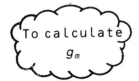

To calculate g_m

In the *Principia*, Newton used geometric arguments to calculate the acceleration of the moon.

The moon describes an approximately circular orbit about the Earth and Newton regarded the moon as a body 'falling towards the Earth' under the action of gravity – just like any other body, such as an apple falling from a tree. If the moon falls a distance S metres in t seconds then its acceleration g_m is given by:

$$S = \tfrac{1}{2} g_m t^2 \qquad (1.3)$$

Newton argued that in 1 second the moon rotates through a small angle θ about the centre of the Earth, O, from P to Q.

In the absence of a gravitational force, the moon would 'continue in a state of uniform motion' along the tangent PT. But the effect of the Earth's gravity is to cause it to fall a distance $TQ = S$. Standard results from trigonometry and calculus give

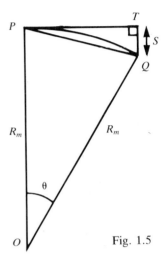

Fig. 1.5

$$\angle TPQ = \frac{1}{2}\theta; \qquad S = PQ \sin \frac{1}{2}\theta$$

$$PQ = 2R_m \sin \frac{1}{2}\theta$$

and therefore

$$S = 2R_m \sin^2 \frac{1}{2}\theta$$

In the limit as $\theta \to 0$, $S = \dfrac{1}{2} R_m \theta^2$ and equation (1.3) gives

$$g_m = R_m \theta^2$$

Now, R_m, the radius of the moon's orbit, is approximately 3.84×10^8 m and its period is 27.3 days $= 2.36 \times 10^6$ s.

Therefore

$$\theta = \frac{2\pi}{2.36 \times 10^6}; \qquad g_m = 2.72 \times 10^{-3} \text{ m s}^{-2}$$

and

$$\frac{g_m}{g} = 2.78 \times 10^{-4} \simeq \frac{1}{3600} \qquad (1.4)$$

As validation for an inverse square law of gravitation, which could be applied at all distances from the Earth's centre, Newton set out to re-calculate g_m/g and compare it with (1.4).

The force on a particle of mass m, at the Earth's surface has magnitude given by:

$$\frac{GMm}{R^2} = mg$$

where M and R are the mass and the radius of the Earth respectively.

When the particle is as far away as the moon

$$\frac{GMm}{R_m^2} = mg_m$$

from which it follows that

$$\frac{g_m}{g} = \left(\frac{R\cdot}{R_m}\right)^2$$

Since the Greeks had shown that the radius of the Earth, R, was approximately 6.4×10^6 m then

$$\frac{g_m}{g} = \frac{1}{3600}$$

which agrees remarkably well with (1.4)! This was a truly astonishing result which we can surmise must have given Newton enormous satisfaction and yet presented him with an additional challenge! On the one hand there is clear validation of an inverse square law; on the other there has been an implicit assumption that an object just a few metres above the Earth's surface is attracted by the Earth as if its whole mass were concentrated at its centre, 6.4×10^6 m below the surface. Newton did in fact establish this result in 1685, twenty years after the publication of the law of gravitation, and the problem is described in detail in chapter 2.

Derivation of Kepler's laws

The problem for Newton was to show how Kepler's observational laws could be derived from his three laws of motion and the inverse square law of gravitation. Recall Kepler's first law:

'The path of every planet is an ellipse, with the Sun at one focus.'

For the moon, Newton had shown that an inverse square law is an appropriate force law to give a **circular** orbit with the Earth at the **centre**. Clearly this is a special case of Kepler's first law. Since Newton's geometrical proof that planetary orbits are elliptical is both long and technically difficult, a modern approach is given here.

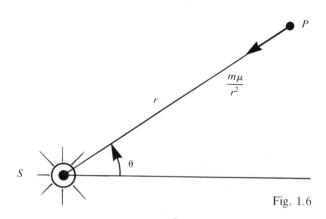

Fig. 1.6

The motion of a planet, of mass m, is described using polar co-ordinates (r, θ) with respect to an origin at the sun. It is acted upon by a gravitational force of magnitude

$$F = \frac{m\mu}{r^2} \quad \text{where} \quad \mu = GM_s$$

(M_s is the mass of the sun) and the radial and tangential components of its acceleration are

$$(\ddot{r} - r\dot{\theta}^2) \quad \text{and} \quad \frac{1}{r}\frac{d}{dt}(r^2\dot{\theta})$$

respectively.

Newton's second law then gives:

$$m(\ddot{r} - r\dot{\theta}^2) = -\frac{m\mu}{r^2} \qquad (1.5)$$

and

$$\frac{m}{r}\frac{d}{dt}(r^2\dot{\theta}) = 0 \qquad (1.6)$$

where (1.6) integrates to give

$$r^2\dot{\theta} = \text{constant} = h. \qquad (1.7)$$

This states that the angular momentum per unit mass is constant and equation (1.5) now becomes

$$\ddot{r} - \frac{h^2}{r^3} = -\frac{\mu}{r^2} \qquad (1.8)$$

This equation is then simplified by introducing $u = \dfrac{1}{r}$,

$$\dot{r} = -\frac{1}{u^2}\dot{u} = -\frac{1}{u^2}\frac{du}{d\theta}\dot{\theta} \text{ and equation (1.7) gives}$$

$$\dot{r} = -h\frac{du}{d\theta}$$

Differentiating again

$$\ddot{r} = -h\frac{d^2u}{dt\,d\theta} = -h\frac{d^2u}{d\theta^2}\dot{\theta} = -h^2u^2\frac{d^2u}{d\theta^2}$$

and substituting into equation (1.8) yields

$$\frac{d^2u}{d\theta^2} + u = \frac{\mu}{h^2}$$

This is a forced simple harmonic motion equation with particular solution $\frac{\mu}{h^2}$ and complementary function $u = A\cos(\theta + \epsilon)$ where A and ϵ are abitrary constants. The general solution is therefore

$$u = A\cos(\theta + \epsilon) + \frac{\mu}{h^2}$$

or

$$r = \frac{\dfrac{h^2}{\mu}}{\left(1 + \dfrac{Ah^2}{\mu}\cos(\theta + \epsilon)\right)} \qquad (1.9)$$

Interpretation

Equation (1.9) may be recognised as the equation of a conic

$$r = \frac{l}{1 + e\cos\theta}$$

with semi-latus rectum

$$l = \frac{h^2}{\mu} \qquad (1.10)$$

and eccentricity $e = \dfrac{Ah^2}{\mu}$. l, e and ϵ are determined from two initial conditions, the initial velocity and position. The conic is an ellipse, hyperbola or parabola according as $e < 1$, $e > 1$ or $e = 1$. Intuitively, if the initial speed of a planet is small (large), then its angular momentum, h, is also small (large) such that eccentricity, e, is less than 1(greater than 1) and the planet orbits the sun in an ellipse (hyperbola).

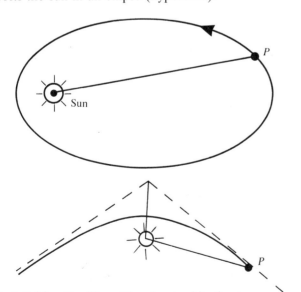

Fig. 1.7 (a) $e<1$: ellipse. The planet orbits the sun.
(b) $e>1$: hyperbola. The planet escapes from the sun.

The critical case $e = 1$ arises when

$$\frac{Ah^2}{\mu} = 1$$

and corresponds to a parabolic orbit. An interesting question arises: 'Why don't some planets orbit in hyperbolas?' and the answer must be 'Perhaps they did!'

Now recall Kepler's second law:

'The line joining the Sun to a planet sweeps out equal areas in equal "intervals of time".'

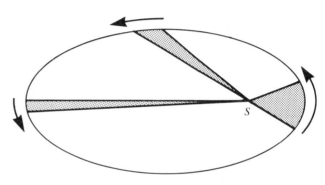

Fig. 1.8

This follows using calculus by considering the motion of a planet from P to P' in a time δt, during which it sweeps out an area δA

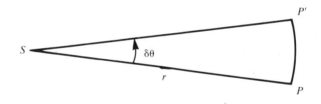

Fig. 1.9

$$\delta A = \tfrac{1}{2}r^2 \cdot \delta\theta$$

$$= \tfrac{1}{2}r^2\,\dot\theta \cdot \delta t$$

Therefore

$$\dot A = \lim_{\delta t \to 0} \frac{\delta A}{\delta t} = \tfrac{1}{2}r^2\dot\theta = \tfrac{1}{2}h$$

which from (1.7) is a constant. As a consequence of this result, and from fig. 1.8, a planet must be moving faster when near the sun than when far away.

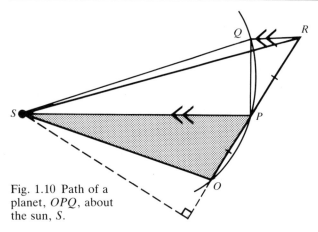

Fig. 1.10 Path of a planet, OPQ, about the sun, S.

Newton's approach to this problem was, however, purely geometrical and made use of the elementary theorem that triangles with the same base and height have equal areas.

Consider a planet orbiting the sun along the arc OPQ and subject only to a central force directed towards S. Initially the planet is at O and after 1 unit of time it is at P such that $\overrightarrow{OP}$ represents its average velocity. After a further unit of time the planet is at Q such that $\overrightarrow{PQ}$ represents its new average velocity.

If $PR = OP$ then the change in velocity, $\delta\mathbf{v} = \overrightarrow{RQ}$

Newton's second law implies that $\overrightarrow{RQ} \parallel \overrightarrow{PS}$.

Geometrically it follows that:

area of $\triangle SOP$ = area of $\triangle SPR$ (same height, equal length base)
area of $\triangle SPR$ = area of $\triangle SPQ$ (same height, equal length base)
area of $\triangle SOP$ = area of $\triangle SPQ$: **Kepler's equal areas rule**.

Recall Kepler's third law:

'The square of a planet's period, T, is proportional to the cube of the semi-major axis of the ellipse, a.'

$$T^2 \propto a^3$$

The time period for orbiting the sun is given by

$$T = \frac{\text{Area to be swept out}}{\text{Rate at which area is swept out}} = \frac{\pi ab}{\dfrac{h}{2}}$$

where a and b are the lengths of the semi-major and minor axes respectively. In fact the geometry of conic sections gives $b^2 = a^2(1 - e^2)$ and $l = a(1 - e)$ and from (1.10)

$$l = \frac{h^2}{\mu}$$

Therefore

$$T^2 = \frac{4\pi^2 a^2 b^2}{h^2} = \frac{4\pi^2 a^4 (1 - e^2)}{l\mu}$$

giving

$$T^2 \propto a^3$$

Note that the special case of a circular orbit is more elementary.

The central force has magnitude $F = \dfrac{m\mu}{r^2}$ and the acceleration is $r\omega^2$.

Newton's second law then gives

$$\frac{m\mu}{r^2} = mr\omega^2$$

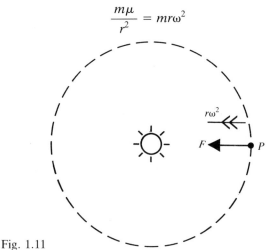

Fig. 1.11

and therefore

$$r^3\omega^2 = \text{constant}$$

Since

$$T = \frac{2\pi}{\omega}$$

then

$$T^2 \propto r^3.$$

The mass of the planets, their periodic time and eccentricity of their orbits is given in table 1.

Table 1

	Mean radius of orbit $r\,(\times 10^6\,\text{km})$	Period $(\times 10^6\,\text{seconds})$	Eccentricity
Mercury	57.9	7.62	0.206
Venus	108	19.4	0.007
Earth	150	31.6	0.017
Mars	228	59.4	0.093
Jupiter	778	376	0.048
Saturn	1430	932	–
Uranus	2870	2654	0.047
Neptune	4500	5214	–
Pluto	5910	7839	–

1.1(d) Limitations of Newton's laws

By relating force and motion in a simple way, Newtonian mechanics provides a powerful tool for analysing and predicting the motion of objects – from the very large to very small. Nevertheless, like any other physical theory it does have its limitations.

For the analysis of fast moving particles, whose

speed is of the order of the speed of light, modifications to space, time and mass are required as given in Einstein's special theory of relativity. This branch of mechanics is known as **relativistic mechanics**. For the analysis of events taking place over very small distances – on an atomic or subatomic scale – Newtonian mechanics gives way to **quantum mechanics**.

Besides the above limitations there is also the question of frames of reference. 'In which frame or frames of reference are Newton's laws valid?' This question is considered in chapter 2, section 2.

1.2 CONSERVATION LAWS: Momentum and energy

To appreciate Newtonian mechanics is to become aware of its essential coherence and simplicity. Students who fail to recognise this tend to regard the subject as no more than a collection of different laws and principles to be applied to various problems. This arises partly from the stereotyped problems which students are taught to solve. Research in mechanics has shown that the difference between expert and novice problem solvers is that while the former classify problems according to which principles apply, the latter classify them according to superficial appearances. For example, a novice sees a problem as being an inclined plane problem or a vertical circle problem. An expert, on the other hand, will recognise a problem as one where Newton's second law or an energy principle should be applied. Perhaps it is inevitable that novices should think in this way at first. But if they are ever to become experts they must appreciate the principles of mechanics and how they relate to one another.

At the heart of the subject are Newton's laws which relate force and motion. In addition, it is necessary to have a number of force laws; Newton's gravitational law for weight, Hooke's law for tension, Newton's law of friction and so on.

But where do the laws of conservation of momentum and energy fit in? Are they extra laws to be added to Newton's laws?

In the study of mechanics, Newton's laws are central and all the conservation laws can be deduced from them. We shall illustrate this by first considering motion in one dimension and then generalising to two and three dimensions.

1.2(a) Momentum and kinetic energy of a particle

The effect of a force **F** on the motion of a body (modelled as a particle of mass m) can be described in two different ways, using either the concept of momentum or that of kinetic energy. To see this consider Newton's second law

$$\mathbf{F} = \frac{d}{dt}(m\mathbf{v}) \qquad (1.11)$$

which for a particle takes the form

$$\mathbf{F} = m\frac{d\mathbf{v}}{dt} \qquad (1.12)$$

For motion in one dimension only

$$\mathbf{F} = F\mathbf{i}, \ \mathbf{v} = v\mathbf{i},$$

$$\frac{dv}{dt} = \frac{dv}{dx} \times \frac{dx}{dt} = v\frac{dv}{dx}$$

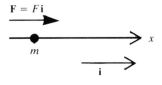

and (1.12) becomes

$$F = mv\frac{dv}{dx} = \frac{d}{dx}(\tfrac{1}{2}mv^2) \qquad (1.13)$$

$$\text{Momentum} = m\mathbf{v}$$

$$\text{KE} = \tfrac{1}{2}mv^2$$

Since $m\mathbf{v}$ represents the momentum of the particle and $\tfrac{1}{2}mv^2$ its kinetic energy (KE), equations (1.11) and (1.13) illustrate how force can be thought of in two different ways

- as that which causes a change in momentum (with respect to time), and
- as that which causes a change in kinetic energy (with respect to distance).

On integrating these equations with respect to time and distance respectively, we obtain

$$\int F\mathbf{i} \cdot dt = [m\mathbf{v}]_U^V = mV\mathbf{i} - mU\mathbf{i} = \text{the } \textbf{impulse} \text{ of } \mathbf{F} \qquad (1.14)$$

$$\int F \cdot dx = [\tfrac{1}{2}mv^2]_u^v = \tfrac{1}{2}mV^2 - \tfrac{1}{2}mU^2 \qquad (1.15)$$

$$= \text{the } \textbf{work done} \text{ by } \mathbf{F}$$

where the speed of the particle changes fron U to V under the action of **F**. Equations (1.14) and (1.15) describe 'the impulse of a force' and the 'work done by a force' respectively. As an example we can see how momentum and kinetic energy changes when a mass of 1 kg, initially at rest, is acted upon by a constant force of 1 N for a time of 2 seconds. It follows from equation (1.14) that the particle will increase its momentum by 1 kg m s^{-1} in each of the first two seconds:

from 0 to 1 kg m s^{-1} in the 1st second
and from 1 to 2 kg m s^{-1} in the 2nd second.

$F = 1, m = 1$.

1st second : increase in KE $= \frac{1}{2}$	2nd second : increase in KE $= 2 - \frac{1}{2} = \frac{3}{2}$

Though the increase in momentum is the same in each second, the work done by the force and therefore the increase in kinetic energy is not the same:

in the 1st second, $\frac{1}{2}mV^2 - \frac{1}{2}mU^2 = \frac{1}{2}$
and in the 2nd second, $\frac{1}{2}mV^2 - \frac{1}{2}mU^2 = \frac{3}{2}$.

The work done by a force in a time interval δt clearly depends on the displacement of the particle δx and so relates to the speed of the particle v. Mathematically these quantities are related by

$$\delta x = v\delta t$$

and the total work done is given by
$$\int F dx = \int F v dt$$

where Fv is the rate at which the force does work, i.e. the power. An alternative form of the work done – kinetic energy equation (1.15) is therefore

$$\int F v dt = \frac{1}{2}mV^2 - \frac{1}{2}mU^2.$$

It is the **integration** of Newton's second law which provides the concepts of impulse, work done and changes in momentum and kinetic energy. These are at the heart of the conservation laws. To fully understand them we need to break out of the restriction to one dimensional motion. A crucial difference between momentum and energy will then become apparent:

- momentum is a vector quantity,
- energy is a scalar quantity.

1.2(b) Conservation of momentum

For one particle
If there is no external force on a particle, $\mathbf{F} = \mathbf{0}$, then it can be deduced immediately from Newton's second law:

$$\frac{d}{dt}(m\mathbf{v}) = \mathbf{0} \quad \text{or} \quad m\mathbf{v} = \text{constant}.$$

Therefore if there is no external force on a particle, its momentum is conserved.

This is the law of conservation of momentum for a particle. It is also Newton's first law!

For two particles
If two particles of mass m_1 and m_2 are moving with velocities $\mathbf{v}_1$ and $\mathbf{v}_2$ subject to no **external** force, then we wish to know if their total momentum is conserved

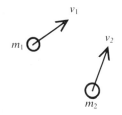

$$m_1\mathbf{v}_1 + m_2\mathbf{v}_2 = \text{constant}?$$

If the particles do not interact then both will move with constant velocity and their total momentum will be conserved. If, however, the two particles do interact (by way of a collision perhaps), there will arise forces of interaction which by Newton's third law are equal and opposite, $\mathbf{R}$ and $-\mathbf{R}$ respectively.

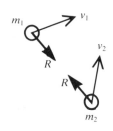

By Newton's second law these forces change the momenta of the two particles

$$\mathbf{R} = \frac{d}{dt}(m_1\mathbf{v}_1) \quad \text{and} \quad -\mathbf{R} = \frac{d}{dt}(m_2\mathbf{v}_2) \qquad (1.16)$$

which when added together give

$$\frac{d}{dt}(m_1\mathbf{v}_1) + \frac{d}{dt}(m_2\mathbf{v}_2) = 0$$

and

$$m_1\mathbf{v}_1 + m_2\mathbf{v}_2 = \text{constant}.$$

We can see that this result follows as a direct consequence of the interaction forces being equal and opposite. From equation (1.16) the change in momentum of one particle is opposite to the change in momentum of the other. The interaction forces **exchange** momentum but cannot **create** any momentum.

An example of this occurs in Newton's cradle, which Newton quotes as evidence for his third law. When the

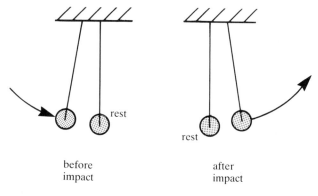

before impact	after impact

See 'Newton's cradle' (Chapter 11).

Fig. 1.12

10

spheres interact, through collision, momentum is passed from one to the other. In fig. 1.12 the sphere on the left gives up its momentum to the other sphere. But this is not generally the case with interactions. If a railway engine collides with a railway carriage and the two link together after impact then clearly the engine gives only a part of its momentum to the carriage.

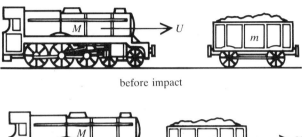

before impact

after impact

Fig. 1.13

Momentum conservation gives

$$M\mathbf{U} = (M + m)\,\mathbf{V}.$$

A vital feature of momentum is that it is a **vector** quantity and consequently momenta in opposite directions may cancel out. (This is in sharp contrast to kinetic energy, as we shall see.)

As an example, consider a cannon, of mass M, which fires a shot of mass m.

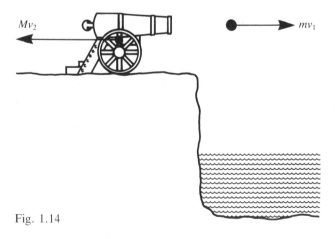

Fig. 1.14

Initially the total momentum is zero and therefore conservation of momentum gives

$$M\mathbf{v}_2 + m\mathbf{v}_1 = 0$$

where $\mathbf{v}_2$ and $\mathbf{v}_1$ are the velocities of the cannon and shot immediately after firing. Clearly the cannon must recoil, as shown, in order that the total momentum remains zero.

Similarly the propulsion of a spaceship in deep space depends on momentum conservation. The spaceship exchanges momentum with its own fuel.

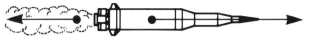

Fig. 1.15

So far we have considered conservation of momentum for two particles, yet the argument may be extended to include any number of particles. The principle of the cancelling out of the forces of interaction is applied to obtain, for a system of N particles,

$$m_1\mathbf{v}_1 + m_2\mathbf{v}_2 + \ldots + m_N\mathbf{v}_N = \text{constant}$$

Going one step further we shall consider a rigid body. This is a collection of particles of mass m_1, $m_2, \ldots, m_N$ held together by forces of interaction in such a way that the distance between any two particles remains constant in time. Individual particles may change their momentum but the total momentum of the body will be conserved (in the absence of external force). Take, for example, a rolling snooker ball, of mass M, and assume there is no external horizontal force. Let the centre of mass move with uniform velocity, say, $\underline{\mathbf{v}}$.

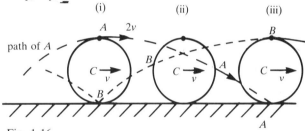

Fig. 1.16

Particles A and B are continually changing their momentum; compare A's velocity for instance in positions (i), (ii) and (iii). But our argument says that the total momentum of the ball is conserved, that is, $M\mathbf{v}$ = constant. Newton's first law is therefore established for a **rigid body** from the three laws for a particle!

> In the absence of an external force, the momentum of a rigid body is constant.
> $$M\mathbf{v} = \text{constant}.$$

It is not difficult to generalise the argument to obtain Newton's second law for a system of particles and also for a rigid body. Consider two particles of mass m_1 and m_2 subject to external forces $\mathbf{F}_1$ and $\mathbf{F}_2$ respectively. If $\mathbf{R}$ and $-\mathbf{R}$ are the forces of interaction then Newton's second law gives

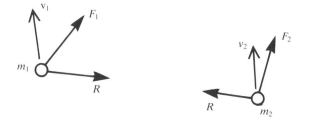

$$\mathbf{F}_1 + \mathbf{R} = \frac{d}{dt}(m_1\mathbf{v}_1)$$

and

$$\mathbf{F}_2 - \mathbf{R} = \frac{d}{dt}(m_2\mathbf{v}_2)$$

which implies that

$$\mathbf{F}_1 + \mathbf{F}_2 = \frac{d}{dt}(m_1\mathbf{v}_1) + \frac{d}{dt}(m_2\mathbf{v}_2)$$

writing $\mathbf{F}_1 + \mathbf{F}_2 = \mathbf{F}$, the resultant external force, then

$$\mathbf{F} = \frac{d}{dt}(m_1\mathbf{v}_1 + m_2\mathbf{v}_2) \qquad (1.17)$$

> The resultant external force is equal to the rate of change of total momentum of the system.

In fact the centre of mass, C, of a system of particles is defined in such a way that, for two particles,

$$(m_1 + m_2)\,\mathbf{v}_C = m_1\mathbf{v}_1 + m_2\mathbf{v}_2$$

where $\mathbf{v}_C$ is the velocity of the centre of mass, C.

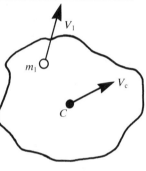

Fig. 1.17

For N particles

$$(\Sigma m_i)\,\mathbf{v}_C = \Sigma m_i\mathbf{v}_i \qquad (1.18)$$

Equation (1.17) then becomes Newton's second law for the two particle system,

$$\mathbf{F} = \frac{d}{dt}(m\mathbf{v}_C) \qquad (1.19)$$

For a system of N particles, or a rigid body, equation (1.17) can be generalised such that together with equation (1.18) it reduces to (1.19).

Newton's second law is therefore established for a system of particles and a rigid body, where the velocity of the system or body is taken to be the velocity of the **centre of mass** of the system or body. For example, it is the **centre of mass** of the Earth–moon system (and not the Earth) which orbits the sun.

1.2(c) The work done – kinetic energy equation

The work done – kinetic energy equation, (1.15), can be extended to two dimensions by considering a force

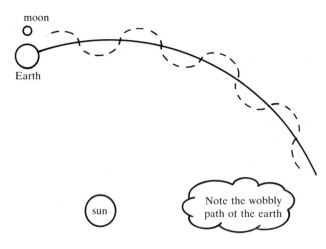

Fig. 1.18

$\mathbf{F} = F_x\mathbf{i} + F_y\mathbf{j}$ acting on a particle of mass m, whose velocity at time t is $\mathbf{V} = u\mathbf{i} + v\mathbf{j}$.

Newton's second law gives two component equations in the x and y directions respectively

$$F_x = m\,\frac{du}{dt} = mu\,\frac{du}{dx}$$

$$F_y = m\,\frac{dv}{dt} = mv\,\frac{dv}{dy}$$

which can be integrated as the particle moves from A, with position vector $\mathbf{r}_1$ and velocity $\mathbf{V}_1 = (u_1,v_1)$, to a position B, with position vector $\mathbf{r}_2$ and velocity $\mathbf{V}_2 = (u_2,v_2)$.

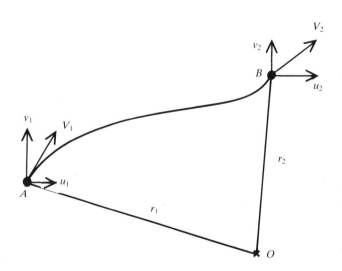

Fig. 1.19

In particular, equation (1.15) gives

$$\int_A^B F_x\, dx = \tfrac{1}{2}m\, u_2^2 - \tfrac{1}{2}m\, u_1^2$$

$$\int_A^B F_y\, dy = \tfrac{1}{2}m\, v_2^2 - \tfrac{1}{2}m\, v_1^2$$

Therefore

$$\int_A^B F_x\, dx + \int_A^B F_y\, dy = \tfrac{1}{2}m\, V_2^2 - \tfrac{1}{2}m\, V_1^2$$

and the left-hand side is precisely $\int_A^B \mathbf{F} \cdot d\mathbf{r}$, the work done by the force $\mathbf{F}$ in moving the particle from A to B.

$$\int_{\mathbf{r_1}}^{\mathbf{r_2}} \mathbf{F} \cdot d\mathbf{r} = \tfrac{1}{2}m\, V_2^2 - \tfrac{1}{2}m\, V_1^2 . \qquad (1.20)$$

This is the general form of the work done – kinetic energy equation.

Observations

1 For a constant force $\mathbf{F}$, the work done is given by

$$\int_{\mathbf{r_1}}^{\mathbf{r_2}} \mathbf{F} \cdot d\mathbf{r} = \mathbf{F} \cdot \int_{\mathbf{r_1}}^{\mathbf{r_2}} d\mathbf{r} = \mathbf{F} \cdot (\mathbf{r}_2 - \mathbf{r}_1)$$

$$= F\, d \cos \theta \text{ where } d = |\mathbf{r}_2 - \mathbf{r}_1|.$$

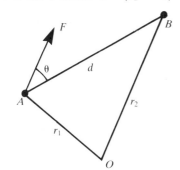

Fig. 1.20

2 The work done – kinetic energy equation, (1.20), has exactly the same form as the equation for motion in one dimension (1.15). Extension to three dimensions is straightforward.
3 An alternative derivation of equation (1.20) is to take the scalar product of Newton's second law, with velocity $\mathbf{V}$, and integrate with respect to time.

$$\mathbf{F} \cdot \mathbf{V} = m \mathbf{V} \cdot \frac{d\mathbf{V}}{dt} = \frac{d}{dt}\left(\tfrac{1}{2}m \mathbf{V} \cdot \mathbf{V}\right) = \frac{d}{dt}\left(\tfrac{1}{2}m V^2\right)$$

therefore

$$\int_{t_1}^{t_2} \mathbf{F} \cdot \mathbf{V}\, dt = \tfrac{1}{2}m V_2^2 - \tfrac{1}{2}m V_1^2 \qquad (1.21)$$

since $d\mathbf{r} = \mathbf{V}\, dt$ then (1.21) is identical to (1.20).
4 The particular form of the integral in equation (1.21) suggests the rate of doing work is now $\mathbf{F} \cdot \mathbf{V}$. An important case occurs when $\mathbf{F} \cdot \mathbf{V} = 0$, i.e. when either $\mathbf{V} = \mathbf{0}$ or $\mathbf{V}$ is perpendicular to $\mathbf{F}$ ($\mathbf{V} \perp \mathbf{F}$).

> If $\mathbf{F} \perp \mathbf{V}$ then $\mathbf{F}$ does no work! If the resultant force $\mathbf{F}$ does no work, kinetic energy is conserved.

Consider two examples.

Example 1
A particle moves in a circle under the action of a force $\mathbf{F}$ acting towards the centre.

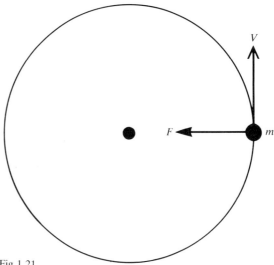

Fig 1.21

The velocity $\mathbf{V}$ is tangential, $\mathbf{F}$ is radial and so $\mathbf{F} \cdot \mathbf{V} = 0$. Therefore $\tfrac{1}{2}mV^2 = \text{constant}$, which implies that $V = \text{constant}$.

We see that speed **is conserved**, since $\mathbf{F}$ does no work but momentum $m\mathbf{V}$ is not conserved, since $\mathbf{F}$ certainly changes momentum.

The difference between momentum $m\mathbf{V}$ as a vector and kinetic energy $\tfrac{1}{2}mV^2$ as a scalar is sharply contrasted here. When $\mathbf{F} \perp \mathbf{V}$ the applied force changes the direction of motion but **not its speed**.

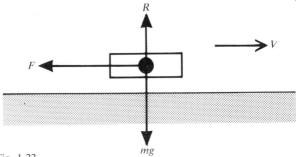

Fig. 1.22

Example 2
Consider the motion of a puck on ice. Forces $\mathbf{R}$ and $m\mathbf{g}$ do no work but the friction force $\mathbf{F}$ is opposite to $\mathbf{V}$ and so does work $-\int FV\, dt$. If we assume the ice to be smooth so that $\mathbf{F} = \mathbf{0}$ then the work done is zero and the KE of the puck is constant, i.e. $V = \text{constant}$. In this case $m\mathbf{V} = \text{constant}$ also since there is no applied force to change momentum.

1.2(d) Conservation of energy for a particle

So far we have not discussed potential energy, PE, or the usual form of the law of conservation of mechanical energy which states that:

> If no external forces (other than gravity) do work on a particle, then the total mechanical energy (KE + PE) of the particle is constant.

Since $\frac{1}{2}mV^2$ represents the KE of a particle having speed V and $\int \mathbf{F} \cdot d\mathbf{r}$ represents the work done by an external force $\mathbf{F}$, it follows from the work–energy equation that:

(a) the work done by the external force on a particle is equal to the change in KE of the particle.

(b) the work needed to reduce a moving particle to rest (i.e. the work done **by the particle** when it is brought to rest) is equal to its KE.

Consequently the KE of a particle is 'stored work', in the sense that it stores the work done by increasing its speed, and yields or gives up that amount of work when it is slowed down again. Consider, for example, when a dart enters a target and comes to rest. The KE is reduced to zero under the action of a resistance force exerted by the target.

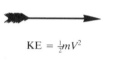

$$\text{KE} = \tfrac{1}{2}mV^2 \qquad\qquad \text{KE} = 0$$

Fig. 1.23

Note, however, that when the KE of a body diminishes, it need not be 'lost', as in the case of the dart, but rather it may be transformed into another form of energy possessed by the body, namely potential energy (PE).

 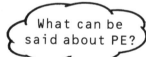

The PE of a particle is the energy that it has by virtue of its position. Compare this with kinetic energy, which a body has by virtue of its speed. If the particle is acted upon by **gravity** then it has PE defined as follows:

> Potential energy is the work done by an external force $\mathbf{G}$ in moving the particle from a reference position $\mathbf{r}_0$ to its current position $\mathbf{r}$ at constant velocity.

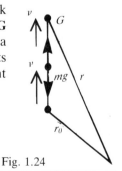

Fig. 1.24

Clearly $\mathbf{G} = -m\mathbf{g}$ so that the total force acting on the particle, $\mathbf{G} + m\mathbf{g}$, is identically zero (which is necessary for it to move with constant velocity).

Therefore

$$\text{PE} = \int_{\mathbf{r}_0}^{\mathbf{r}} -m\mathbf{g} \cdot d\mathbf{r} \qquad (1.22a)$$

or

$$\text{PE} = \int_{\mathbf{r}}^{\mathbf{r}_0} m\mathbf{g} \cdot d\mathbf{r} \qquad (1.22b)$$

Note that

1 PE of a particle at the reference position $\mathbf{r}_0$ is zero;

$$\int_{\mathbf{r}_0}^{\mathbf{r}_0} m\mathbf{g} \cdot d\mathbf{r} = 0 .$$

2 Equation (1.22b) shows that potential energy may also be defined as the work done by gravity in moving the particle from $\mathbf{r}$ to $\mathbf{r}_0$.

3 Taking the surface of the Earth as the zero PE level, a particle of mass m, at height h above the ground, has PE $= mgh$. This follows from the definition of potential energy as the work done by an external force $\mathbf{G} = -m\mathbf{g}$ to take the particle at constant velocity from $y = 0$ to $y = h$.

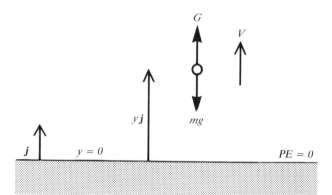

Fig. 1.25 Particle acted upon by external force G and gravity $m\mathbf{g} = -mg\mathbf{j}$

Work done $= \int_{y=0}^{h} \mathbf{G} \cdot d\mathbf{r} = \int_{y=0}^{h} mg\mathbf{j} \cdot d\mathbf{r}$

$\qquad\qquad = \int_0^h mg \, dy = mgh.$

Therefore PE $= mgh$.

4 Kinetic energy can be converted into potential energy and vice versa.

Consider a ball, of mass m, thrown vertically with speed U.

Initially it has KE $= \frac{1}{2}mU^2$ and PE $= 0$.

As it rises, its KE diminishes, eventually becoming zero when it reaches its maximum height h. During this time its PE increases from zero to a maximum value of mgh.

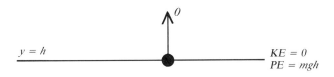

$y = h$ $KE = 0$
                              $PE = mgh$

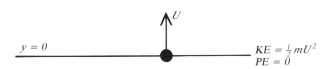

$y = 0$ $KE = \frac{1}{2}mU^2$
                              $PE = 0$

Fig. 1.26

Clearly KE has been converted into PE. When the ball falls back to Earth under the action of $m\mathbf{g}$ the reverse happens as PE is converted into KE.

5 Potential energy (mgh) is the work done by gravity $m\mathbf{g}$ on the particle as it falls back to Earth! (See note 2 above.)

Work done $= \int_{y=h}^{0} m\mathbf{g} \cdot d\mathbf{r} = \int_{y=h}^{0} (-mg\mathbf{j}) \cdot (dy\mathbf{j})$

$\qquad\qquad = mgh$.

> How is the PE of a particle related to its KE?

We shall show that provided gravity is the only force which does work on a particle then the sum of its kinetic and potential energies is constant; KE + PE = constant.

Consider a particle, of mass m, moving under gravity from A (position vector $\mathbf{r}_1$) to B (position vector $\mathbf{r}_2$).

Let the height and velocity at A be h_1 and $\mathbf{V}_1$ and at B be h_2 and $\mathbf{V}_2$. The work–energy equation, equation (1.15) now gives

$$\int_{\mathbf{r}_1}^{\mathbf{r}_2} m\mathbf{g} \cdot d\mathbf{r} = \tfrac{1}{2}mV_2^2 - \tfrac{1}{2}mV_1^2$$

Since $\qquad\qquad m\mathbf{g} \cdot d\mathbf{r} = (-mg\mathbf{j}) \cdot dy\mathbf{j}$ then

$$-\int_{h_1}^{h_2} mg\, dy = \tfrac{1}{2}mV_2^2 - \tfrac{1}{2}mV_1^2$$

$$-mg(h_2 - h_1) = \tfrac{1}{2}mV_2^2 - \tfrac{1}{2}mV_1^2$$

therefore

$$\tfrac{1}{2}mV_1^2 + mgh_1 = \tfrac{1}{2}mV_2^2 + mgh_2$$

$$\tfrac{1}{2}mV^2 + mgh = \text{constant}$$

or

$$\boxed{\text{KE} + \text{PE} = \text{constant.}} \qquad (1.23)$$

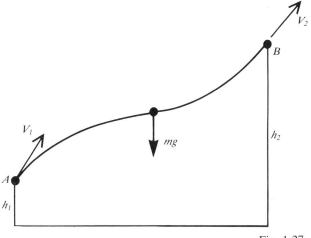

Fig. 1.27

Notes

Note the following two points:

(a) $m\mathbf{g}$ must be the **only force which does work** on the particle.

(b) the actual path from A to B is irrelevant. Since $m\mathbf{g}$ is a constant force,

$$\text{work done} = \int_{\mathbf{r}_1}^{\mathbf{r}_2} m\mathbf{g} \cdot d\mathbf{r} = mg(h_2 - h_1)$$

and depends only on the end points A and B where h is equal to h_1 and h_2 respectively, and not on the route taken between A and B. This is why gravity is called a **conservative force**.

The following two examples illustrate how useful the energy principle can be in practical applications.

Example 1

A ball moves on a smooth track from A to B. The speed V_B can be found using the conservation of energy. Equation (1.23) applies, because $\mathbf{N} \perp \mathbf{V}$ and $m\mathbf{g}$ is the only force that does work.

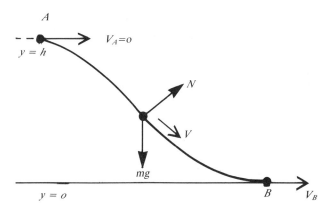

Fig. 1.28   *See 'Loop the loop (i)' (chapter 8)*

Energy at A = energy at B,

so $\qquad\qquad 0 + mgh = \tfrac{1}{2}mV_B^2$

$$V_B = \sqrt{2gh} \ .$$

Note that the shape of the track is irrelevant. The crucial assumption is that there is no friction force.

Example 2 A simple pendulum

Tension $\mathbf{T} \perp \mathbf{V}$. The only force doing work is $m\mathbf{g}$ and therefore the energy equation applies.

See 'Simple pendulum' (chapter 10)

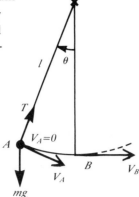

Fig. 1.29

Energy at A = energy at B

$$\tfrac{1}{2}m\cdot 0^2 + mgl(1 - \cos\theta) = \tfrac{1}{2}mV_B^2 + 0$$

and so

$$V_B^2 = 2gl(1 - \cos\theta).$$

1.2(e) Energy of systems of particles

Earlier we saw how the conservation of momentum equation could be extended to both systems of particles and rigid bodies since the interaction forces only **exchange** momentum between the interacting particles. Unfortunately the law of conservation of energy cannot always be extended to systems of interacting particles. This is readily seen by considering the two examples described in section 1.2(b).

1 Newton's cradle

We considered the impact of two spheres in which momentum is transferred from one to the other. In this case momentum is conserved and kinetic energy is also conserved. Such a collision is called an **elastic collision**. (Strings and springs which extend and compress with negligible energy loss are similarly called elastic.)

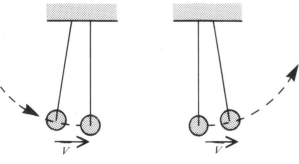

See 'Newtons cradle' (chapter 11)

Fig. 1.30

2 The impact of two railway trucks in which both have a common speed V after collision. If we assume that the two trucks have equal mass m, then $V = \tfrac{1}{2}U$ where U is the speed of one truck before impact.

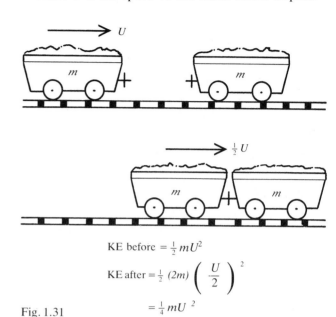

$$\text{KE before} = \tfrac{1}{2}mU^2$$

$$\text{KE after} = \tfrac{1}{2}(2m)\left(\frac{U}{2}\right)^2$$

$$= \tfrac{1}{4}mU^2$$

Fig. 1.31

Such a collision involves a loss of kinetic energy and is called an **inelastic collision**. In general, for collisions of this type, Newton's law of restitution provides a crude model which relates relative velocities before and after impact.

Newton's law of restitution

$$(V_2 - V_1) = -e(U_2 - U_1) \qquad (1.24)$$

The relative speed of two bodies after impact is e times their relative speed before impact and e, the **coefficient of restitution**, is a number between 0 and 1.

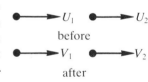

The inelastic collision above arises when $e = 0$, so that

$$V_2 = V_1.$$

Consider a final case of an inelastic collision in which all the kinetic energy is lost at impact.

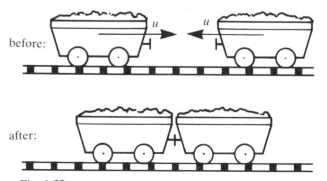

Fig. 1.32

The carriages collide and link. Conservation of momentum dictates that the momentum after collision is zero. Therefore $V = 0$ and the kinetic energy is also zero.

Conclusion: conservation of energy cannot generally be applied to interacting systems of particles.

Why is this? It is because the internal forces, even though equal and opposite, can do work on the particles which does not cancel out. The force **vectors** would cancel, but the scalars $\mathbf{F} \cdot \mathbf{V}$ do not (necessarily). Consider the explosion of a pair of particles (shrapnel from a grenade).

The interaction forces $\mathbf{F}$ and $-\mathbf{F}$ cause the particles to move apart with opposite momenta, but the work done by the forces (each of which is in the direction of motion of each piece of shrapnel) does not sum to zero!

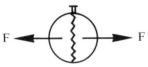

1.2(f) Energy of a rigid body

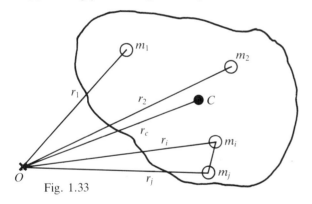

Fig. 1.33

A rigid body can be regarded as a set of N particles, of mass m_i ($i = 1, 2, \ldots, N$) and position vector $\mathbf{r}_i$ *with respect to a fixed origin O, such that the distance between **any** two remains constant in time:*

$$|\mathbf{r}_i - \mathbf{r}_j| = \text{constant}$$

where $i, j \in \{1, 2, \ldots, N\}$.

The mass of the rigid body is given by

$$M = \sum_{i=1}^{N} m_i$$

and its centre of mass, C, will be a point fixed in the rigid body, having position vector $\mathbf{r}_c$ such that

$$M\mathbf{r}_c = \sum_{i=1}^{N} m_i \mathbf{r}_i \qquad (1.25)$$

This equation when differentiated gives (1.18) which was used to give the velocity of the centre of mass.

Potential energy due to gravity

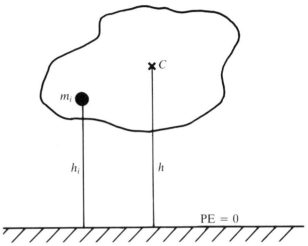

Fig. 1.34 PE of a rigid body $= Mgh$.

If a rigid body has its mass centre C, at a height h above the zero PE level ($y = 0$) then the potential energy of the body is Mgh.

This follows by considering the particle, of mass m_i and height h_i above $y = 0$. Its PE is $m_i g h_i$ and therefore for the rigid body

$$\text{PE} = \sum_{i=1}^{N} m_i g h_i \qquad (1.26)$$

The y component of equation (1.25) gives

$$Mh = \sum_{i=1}^{N} m_i h_i$$

and (1.26) becomes

$$\text{PE} = Mgh$$

This result tells us that the PE of a rigid body of mass M is equivalent to that of a **single particle** of mass M located at C.

Kinetic energy of a rigid body

If the ith particle of a rigid body has mass m_i and velocity $\mathbf{V}_i$ then its KE is $\frac{1}{2}m_i V_i^2$ and therefore the total KE is given by

$$\text{KE} = \sum_{i=1}^{N} \frac{1}{2} m_i V_i^2$$

17

Examples

1 A ball or shot-put moving through the air without rotation, or indeed a puck moving along the ice without rotation.

   In either case the motion is a pure translation; every particle has the same V and the kinetic energy is given by:

$$KE = \sum_{i=1}^{N} \tfrac{1}{2} m_i V^2 \qquad (1.27)$$

$KE = \tfrac{1}{2} MV^2$ if there is no rotation.

2 A disc or catherine wheel rotating about an axis through its mass centre C, with angular speed ω. Here C is at rest and the motion is a pure rotation about C.

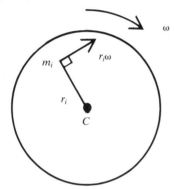

If we consider the ith particle, of mass m_i and distance r_i from C, then it moves in a circle about C with speed $r_i\omega$. Its kinetic energy is $\tfrac{1}{2} m_i r_i^2 \omega^2$ and therefore for the disc

$$KE = \tfrac{1}{2} \sum_{i=1}^{N} m_i r_i^2 \omega^2$$

or

$$KE = \tfrac{1}{2} I \omega^2 \qquad (1.28)$$

where

$$I = \sum_{i=1}^{N} m_i r_i^2$$

is called the moment of inertia of the disc about the axis through C.

In general a rigid body moves in such a way that its motion can be regarded as a 'translation plus a rotation' i.e. a translation of its mass centre C, together with a rotation about C. For example when a snooker ball rolls across the table from A to B, its centre C, is translated and there is rotation about C.

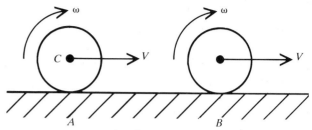

Fig. 1.35 Energy $= \tfrac{1}{2}MV^2 + \tfrac{1}{2}I\omega^2$

It can be shown that the kinetic energy of a rigid body, of mass M, consists of two components

$$KE = \tfrac{1}{2} MV^2 + \tfrac{1}{2} I\omega^2 \qquad (1.29)$$

where the first term represents translational kinetic energy:

> $\tfrac{1}{2}MV^2$ = KE of a single particle of mass M moving with the speed of C.

and the second term represents rotational kinetic energy:

> $\tfrac{1}{2}I\omega^2$ = KE of rotational motion relative to C.

Notes

(a) When there is no rotation, (1.29) reduces to (1.27) – a pure translation.

(b) when there is no motion of the mass centre (1.29) reduces to (1.28) – pure rotation about C.

(c) For a snooker ball, of radius a, rolling along a horizontal table with constant speed V, its angular speed ω, is given by

$$V = a\omega$$

and therefore its kinetic energy is given by

$$KE = \tfrac{1}{2}MV^2 + \tfrac{1}{2}(\tfrac{2}{5}Ma^2)\omega^2$$

$$KE = \tfrac{7}{10}MV^2 \qquad (1.30)$$

where $\tfrac{2}{5}Ma^2$ is the moment of inertia of a sphere about a diameter.

Equation (1.30) provides an important result which illustrates that the rotational KE of either a rolling snooker ball or marble amounts to $\tfrac{2}{7}$ of the total KE regardless of how large or small is the radius. Modelling a rolling marble as a smooth particle is clearly not very accurate; it will involve an error of about 30%.

Energy conservation

If gravity is the only applied force which does work on a rigid body then it can be shown that there is conservation of mechanical energy,

$$KE + PE = \text{constant}$$

Examples

1 A marble rolls on a rough track such that at time t, its height is h and the speed of its mass centre is V.

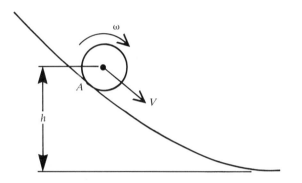

See 'Loop the loop (2)' (chapter 9)   Fig. 1.36

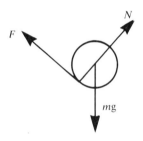

The forces acting on the marble are gravity ($m\mathbf{g}$), friction ($\mathbf{F}$) and normal reaction ($\mathbf{R}$). As we have seen earlier for a particle, $\mathbf{R}$ does no work since $\mathbf{R}$ is perpendicular to $\mathbf{V}_A$ ($\mathbf{R} \cdot \mathbf{V}_A = 0$) where $\mathbf{V}_A$ is the velocity of the point of contact, A, of the marble with the track. Furthermore, for a pure rolling motion, friction does no work; $\mathbf{F} \cdot \mathbf{V}_A = \mathbf{0}$ since $\mathbf{V}_A = \mathbf{0}$ – the point of contact is instantaneously at rest. The energy equation therefore applies and gives

$$\tfrac{7}{10}MV^2 + mgh = \text{constant}.$$

2 A pendulum, consisting of a uniform rod of mass M and length $2a$, rotates about one end where it is smoothly hinged.

If, at time t, the rod makes an angle θ with the downward vertical, then its instantaneous angular speed is $\omega = \dot{\theta}$.

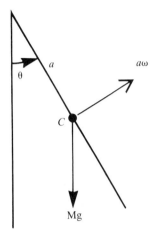

Fig. 1.37

See 'Compound pendulum' (chapter 10)

The speed of C is $a\omega$ and so

$$\text{kinetic energy KE} = \tfrac{1}{2}M\,(a^2\omega^2) + \tfrac{1}{2}I\omega^2$$

where the moment of inertia of a rod about an axis through its centre is given by

$$I = \tfrac{1}{3}Ma^2.$$

potential energy

$$PE = -Mga\cos\theta$$

and the conservation of energy gives

$$\tfrac{2}{3}Ma^2\omega^2 - Mga\cos\theta = \text{constant}.$$

2
MODELLING

Scientists construct models in order to simplify the real world and represent it in some sense.

Newton modelled the Earth:

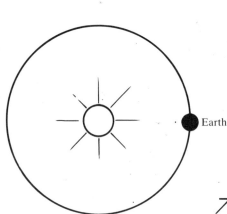

as a **particle** *when considering its orbit around the sun;*

as a **flat plane** *when considering the fall of an apple;*

as a **sphere** *when considering tides and the variation of g with latitude.*

Modelling is at the heart of science.

aims of chapter two

After reading this chapter you will:

- know the three key properties of any model.
- appreciate that Newton modelled the Earth in three different ways when calculating the effect of the Earth's gravity; as a particle, a flat plane and a sphere.
- appreciate that g varies with latitude and height.
- understand that an inertial frame of reference is an idealisation which is used to model actual frames of reference.
- understand the distinction between inertial and non-inertial frames.
- know that linear functions are used to model
  (a) limiting friction
  (b) tension in elastic strings/springs and
  (c) air resistance.

2.1 WHAT IS A MODEL?

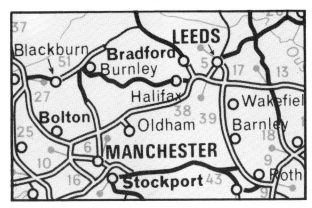

A model is a representation of a real situation. A real situation will invariably contain a rich variety of detail – and any model of it will simplify reality by extracting those features which are considered to be most important.

---

*   **A model is an approximation to reality!**

Example 1

A road map models a real road network and landscape. It provides a two-dimensional representation of a three-dimensional grid. Buildings, mountains and the curvature of the Earth are all ignored.

Example 2

A mathematical model for the time period, T, of the oscillations of a simple pendulum is

$$T = 2\pi \sqrt{\left(\frac{l}{g}\right)} \qquad (2.1)$$

where l is the length of the string, and g is the acceleration due to gravity.

This formula for T is based on a simple model which

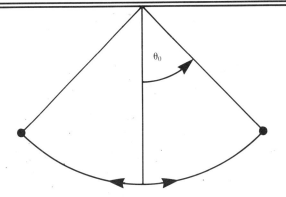

Fig. 2.2 The oscillations of a simple pendulum with amplitude θ_0.

isolates just a few important quantities. Among those ignored are the size and shape of the mass; the diameter, density and elasticity of the string and the density and viscosity of the air. In addition it is assumed that the amplitude of oscillations, θ_0, is 'small'.

Example 3

Newton's laws constitute a model for force and motion. As mentioned earlier (chapter 1) this is a model which Newton used to explain and describe the orbits of the planets and comets, the tides, the variation of g with latitude, the 'precession of the equinoxes' and also, of course, the fall of an apple! Newton's laws are an approximation to reality in the sense that no account is taken of, for example, relativistic effects. Newton assumed that the mass of a particle remained constant whereas Einstein showed that it is dependent on speed, v;

$$m(v) = \frac{m_0}{\left(1 - \dfrac{v^2}{c^2}\right)^{\frac{1}{2}}} \qquad (2.2)$$

where m_0 is the rest mass of the object and c is the speed of light.

By manipulating the model we can hypothetically explore the real situation and make predictions about it. We can ask such questions as 'What happens if. . . ?' and the model will help to provide answers.

Using the map I can plan a route from Leeds to Manchester. Hypothetically, I can journey to Manchester via the M62 or via the A62. The map helps me to predict which will be the shorter of the two routes and which road signs I should see on the way.

The model for the period of oscillations T given by (2.1) predicts a time period of 2 seconds when the length is 1 metre. It will also predict the length of string required for a one second period.

The Newtonian model was used to predict the existence and position of the planets Neptune and Pluto, whose very existence had not previously been suspected. Since then the model has been fundamental to the solution of all problems in science and engineering which involve force and motion.

---

*   **A model is a tool which is only useful when applied to the task for which it was designed!**

One important feature of any model is the set of assumptions upon which it is built; we must never lose sight of these. If the results of a model are applied (and predictions made) in situations where they were not intended, i.e. where assumptions are violated, then the predictions will be inaccurate or even totally invalid.

A map of the London Underground is useful for planning a route on the tube. If it is used to predict the walking time from one place to another it will be unhelpful. I might think 'clearly Stanmore is close to Edgware, so I will get off at Edgware and walk'. Unfortunately, this turns out to be a very long walk indeed, since the map does not represent distances to scale.

Similarly if Newton's laws are applied to the motion of high speed atomic particles (for example electrons moving in an accelerator) then entirely wrong predictions will arise. As mentioned earlier, the Newtonian model fails to predict how inertial mass increases with speed as given by equation (2.2).

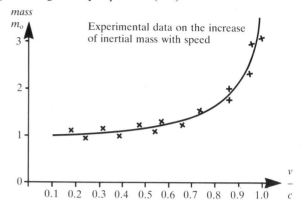

Fig. 2.4 Variation of mass with speed

Equation (2.1) for the time period T is based on a simple model which assumes that the amplitude of oscillations, θ_0, is sufficiently small for $\sin \theta \simeq \theta$. The table below gives the percentage increase in T as amplitude increases:

Amplitude θ_0 (degrees)	% increase in T
0	0.0
15	0.4
30	1.7
45	4.0
60	7.3
90	18.0
135	52.8

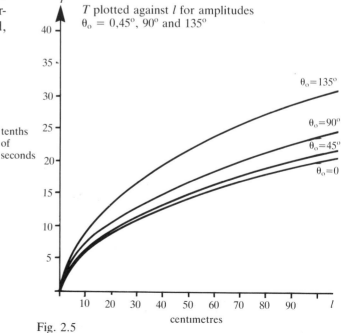

Fig. 2.3 A map section of the London Underground.

Fig. 2.5

Clearly the simple formula for T is good for amplitudes up to about 15° and becomes increasingly inaccurate as amplitude increases beyond about 60°. Additional inaccuracies might also arise if the weight of the bob were small compared to the weight of the string or the force due to air resistance.

Conclusions:
A model is an approximation to reality.
A model enables predictions to be made.
A model is a tool which is only useful when applied to the task for which it was designed.

In general it is essential to understand the limitations of any model and this involves knowing all the assumptions that go into the building of the model. Unfortunately many students of Newtonian mechanics fail to appreciate the nature of many of its basic assumptions. For example:

We assume a body is a particle!
　What is a particle?
We assume a pulley is smooth!
　What is a smooth pulley?
We assume a string is light and inextensible!
　What is a light and inextensible string?

This chapter aims to answer such questions and to investigate such assumptions.

2.2 NEWTON'S MODELLING OF THE EARTH AND ITS GRAVITY

Newton used three laws of motion and the law of gravitation to deduce the motion of various bodies. Each problem required a model for the gravitational force exerted by the Earth (or the sun perhaps) and in order to do this he had to model the Earth, sun and other bodies in mathematical terms. What models did he use?

There are three obvious geometrical models to choose from:

■ the Earth as a particle, that is, a mass concentrated at a point.
■ the Earth as a plane, that is, a flat Earth.
■ the Earth as a sphere.

These are intuitively obvious choices because of the pictures we have in our minds of the Earth from different perspectives.

From the standpoint of the solar system the Earth 'looks' just like a point. (Fig. 2.6(a)). From the top of a tower the 'Earth', particularly a seascape, may look flat. (Fig. 2.6(b)) whereas from an orbiting spacecraft it looks spherical, (Fig. 2.6(c)).

So which did Newton use? The answer is that he used all three! When considering the Earth's orbit

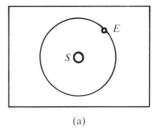 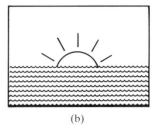

(a)　　　　　　　　　　　(b)

(c)

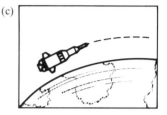

Fig. 2.6 The Earth modelled as a point, as a plane and as a sphere.

around the sun, the Earth is a particle. When considering the fall of an apple from a tree, the Earth is flat. When modelling global phenomena such as the movement of the tides, the Earth is a sphere.

To be more accurate we should say the Earth 'is modelled' by a particle, plane or sphere in each case. Does this make sense? Is it valid to model the Earth in different ways according to the problem? Let us look at each of these cases in turn.

2.2(a) Modelling the Earth (and the sun) as a particle

In the analysis of planetary orbits the sun and Earth are modelled as points. The path of the Earth is a curve. The 'motion' (Newton's word for momentum) of the Earth is represented by a tangent. The change in the 'motion' is caused by gravity acting towards the sun, which is another point mass.

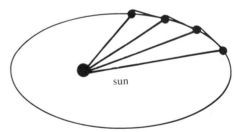

sun

Fig. 2.7 Newton's method of considering changes in the motion of the Earth caused by gravitational attraction from the sun.

Only two bodies are considered here, the effects of other planets are for the moment ignored. The rotation of the Earth about its axis is also ignored. Finally and most importantly, we assume that the gravitational effect of a body is the same as that of a particle of the same mass positioned at its centre of mass. All these assumptions are implicit in the diagrams and derivation of Kepler's laws which Newton provided in the *Principia* (see chapter 1).

Some simple mathematics shows why it is valid to model the sun as a particle provided that its radius r

is small compared to its distance from the Earth d

$$r \ll d.$$

Consider the sun as a collection of small masses m_i, $1 \le i \le N$, whose distance from the centre of the sun is r_i. Then let $\sum_{i=1}^{N} m_i = M_s$ be the mass of the sun and let M be the mass of the Earth.

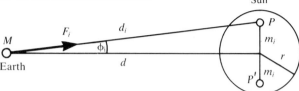

Fig. 2.8 The sun considered as a collection of small masses m_i ($1 \le i \le N$), distance r_i from the centre of the sun and d_i from the Earth.

The gravitational attraction exerted by mass m_i on the earth has magnitude given by

$$F_i = G \frac{M m_i}{d_i^2}$$

and the sum of the components along the line of centres is

$$\sum_{i=1}^{N} F_i \cos \phi_i = F \text{ say.}$$

(By symmetry the component of force normal to the line of centres due to mass m_i at P will be cancelled by an equal and opposite contribution from mass m_i at P'.)

Since $\dfrac{d-r}{d_i} \le \cos \phi_i \le 1$ and $d-r \le d_i \le d+r$

then

$$\frac{GM(d-r)}{(d+r)^3} \Sigma m_i \le F \le \frac{GM}{(d-r)^2} \Sigma m_i$$

therefore

$$\frac{GMM_s}{d^2}\left(1 - \frac{r}{d}\right)\left(1 + \frac{r}{d}\right)^{-3} \le F \le \frac{GMM_s}{d^2}\left(1 - \frac{r}{d}\right)^{-2}.$$

Expanding by the Binomial theorem gives

$$\frac{GMM_s}{d^2}\left(1 - \frac{4r}{d} + \ldots\right) \le F \le \frac{GMM_s}{d^2}\left(1 + \frac{2r}{d} + \ldots\right)$$

Incidentally Newton not only knew of the generalised Binomial theorem, he discovered it. In particular for $|x| < 1$ and real number n,

$$(1 + x)^n = 1 + nx + n(n-1)\frac{x^2}{2!} + n(n-1)(n-2)\frac{x^3}{3!} + \ldots$$

Hence if $r \ll d$ then the particle model for the sun gives a result for the gravitational force of attraction on the Earth in which the error is, at most, of order $\left(\dfrac{r}{d}\right)$.

For the Earth and sun

$$\frac{r}{d} = \frac{6.96 \times 10^8}{1.5 \times 10^{11}} \simeq 0.005.$$

Similarly, for the moon and Earth

$$\frac{R}{d} = \frac{6.4 \times 10^6}{3.8 \times 10^8} \simeq 0.017.$$

First conclusion
When calculating the force of gravity due to the Earth on a body (at a distance d from its centre), the Earth can be modelled as a particle provided $\dfrac{R}{d} \ll 1$ where R is the radius of the Earth.

However, this is not the end of the story. Newton sought to calculate the force due to gravity on a body which is at the **surface** of the Earth, and the relevant questions are:

- does the force experienced by an apple obey the inverse square law?
- can we regard the Earth as a particle even when $d \simeq R$? (Certainly the above argument is not valid in this case!)

In chapter 1 the argument given to justify an inverse square gravitational law assumed that the pull of the Earth **F** on an apple of mass m **was the same** as if the Earth's mass M were concentrated at its centre C.

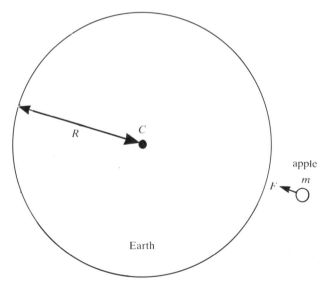

Fig. 2.9 Gravitational attraction between the Earth and an apple.

It is not at all obvious that this is the case. For Newton a key problem was to establish the validity of this result in order to link by a simple law both terrestrial gravity and that experienced by planetary and celestial bodies.

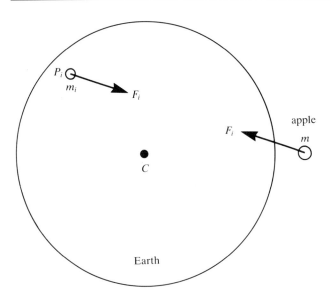

Fig. 2.10 The Earth considered as a collection of particles of mass m_i $(1 < i < N)$

Once again the Earth is regarded as a collection of particles. The particle of mass m_i at P_i attracts the apple with a force $\mathbf{F}_i$ and the sum of all the $\mathbf{F}_i$ $(1 \leq i \leq N)$ will by symmetry act towards C.

Does $\sum\limits_{i=1}^{N} \mathbf{F}_i = \mathbf{F}$?

This problem worried Newton for several years until he derived a proof for the particle model which assumes only that the body is spherically symmetric, that is, it can be regarded as a collection of thin spherical shells of uniform density – rather like the successive layers of an onion. The strategy is to calculate the gravitational force exerted by a thin spherical shell of arbitrary radius, s, and then sum over all shells from $s = 0$ to $s = R$. For this the calculus is needed – and Newton invented that also!

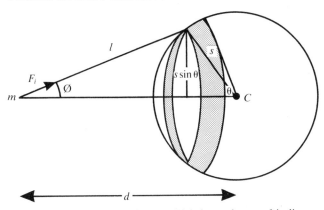

Fig. 2.11 A shell of radius s, which is made up of individual rings.

The diagram shows a shell with centre at C and radius s where $0 \leq s \leq R$. This shell is made up of individual rings like that shown, of radius $s \sin \theta$, width $s d\theta$ and mass $(2\pi s(\sin \theta) s d\theta)\rho = \Delta M$ where $\rho(s)$ is the shell density.

The force of attraction exerted on a mass m by a mass m_i, distance l apart, is given by

$$F_i = \frac{Gmm_i}{l^2}$$

and therefore the total force due to the ring has magnitude

$$\sum F_i \cos \phi = \frac{Gm\Delta M}{l^2} \cos \phi$$

and is directed towards C by symmetry. Summing over all the rings gives the force (of magnitude F_s) due to the shell,

$$F_s = \sum \frac{Gm\Delta M}{l^2} \cos \phi$$

$$= mG2\pi s^2\rho \int_{l=d-s}^{l=d+s} \frac{\sin \theta \cos \phi \, d\theta}{l^2} .$$

Using the cosine rule

$$\cos \phi = \frac{l^2 + d^2 - s^2}{2ld}$$

and

$$\cos \theta = \frac{s^2 + d^2 - l^2}{2sd}; \quad \sin \theta \, d\theta = \frac{ldl}{sd} .$$

Then

$$F_s = \frac{Gm\overline{M}}{2} \int_{d-s}^{d+s} \frac{1}{l^2}\left(\frac{l^2 + d^2 - s^2}{2ld}\right) \cdot \frac{ldl}{sd}$$

and evaluation of the integral yields

$$F_s = \frac{Gm\overline{M}}{d^2}$$

where $\overline{M}$ is the mass of the shell.

Eureka! This result does not involve the radius of the shell and therefore integration over all the spherical shells is straightforward, regardless of how the density $\rho(s)$ varies from the centre of the sphere to the surface

$$F = \frac{GmM}{d^2} .$$

Second conclusion

When calculating the force of gravity due to the Earth on a body (at a distance d from its centre), the Earth can be modelled by a particle provided only that $d \geq R$ and the Earth is assumed to be spherically symmetric.

In conclusion, the Earth is modelled by a particle to determine its orbit about the sun. In addition the particle model (for the Earth) gives the same result as the sphere model in the calculation of the gravitational force on a body close to the surface.

2.2(b) Modelling the Earth as a flat plane

Intuitively, a planet or large sphere can be thought of as approximately flat when we are 'close to its surface'. When an apple falls from a tree or a ball rolls down an inclined plane, the effects due to the earth's curvature can be assumed to be negligible. Consequently gravity acts in the same direction – downwards! Now Galileo showed that all masses close to the Earth's surface fall down with the same acceleration due to gravity, **g**, whose magnitude is g. Therefore we may conclude from Newton's second law that the gravitational force on a mass m of weight W is given by **F** = m**a**, and we have

 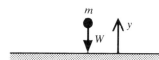

$$W = mg$$

Fig. 2.12

Is this result for weight consistent with the universal law of gravity?

If the body is at a height y above Earth's surface then

$$W = \frac{GmM}{(R + y)^2}$$

$$W = \frac{GmM}{R^2}\left(1 + \frac{y}{R}\right)^{-2}$$

Consistency, therefore, requires that

$$\frac{GM}{R^2} = g$$

and also that $\dfrac{y}{R}$ is negligible. Consequently it is permissible to write mg for the weight of any body which remains close to the Earth, such that its height y is much smaller than the radius of the Earth. As y increases the model $W = mg$ becomes gradually more inaccurate until errors become unacceptable. Similarly the assumption that the **direction** of the gravitational force is constant becomes unacceptable as y increases, as does the assumption that the Earth's curvature is negligible. Consider the paths of long range projectiles. As the range increases, the 'parabolic path' becomes increasingly distorted, eventually becoming elliptical.

Fig. 2.13

(a) flat   (b) curved   (c) sphere

2.2(c) Modelling the Earth as a sphere

In a sense, the sphere model bridges the gap between the particle and flat Earth models. In the diagram, Newton shows in a thought experiment how the parabolic path of a projectile gradually becomes an elliptical orbit as the range increases.

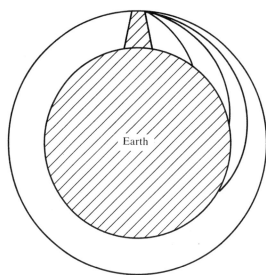

Fig. 2.14 Newton's thought experiment illustrating the transition from a parabolic to an elliptical path.

The model of the Earth as a sphere was used by Newton to explain the lunar tides as a consequence of the gravitational attraction of the moon. He argued as follows: the Earth and moon exert gravitational forces on each other and their centre of mass C lies inside the Earth at approximately 5000 kilometres from the Earth's centre.

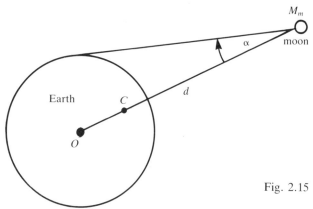

Fig. 2.15

Ignoring the rotation of the Earth about its axis then its centre O has an acceleration a_0 given by Newton's laws:

$$Ma_0 = \frac{GMM_m}{d^2}$$

therefore

$$a_0 = \frac{GM_m}{d^2}$$

where M_m is the mass of the moon and d its distance from the Earth.

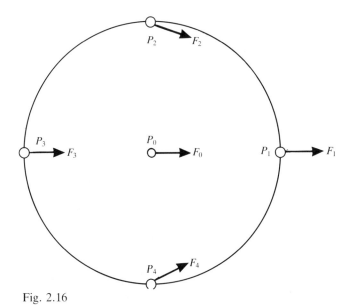

Fig. 2.16

Now consider the Earth as a collection of particles, each of mass m and experiencing the same acceleration a_0. The pull of the moon, however, is different on particles P_0, P_1, P_2, P_3 and P_4 – say $\mathbf{F}_0$, $\mathbf{F}_1$, $\mathbf{F}_2$, $\mathbf{F}_3$ and $\mathbf{F}_4$ as shown in the diagram

For the particle at the Earth's centre $F_0 = ma_0$; the gravitational force has magnitude identically equal to the mass × acceleration. The particle P_1 has a gravitational force of attraction of magnitude F_1 which is greater than F_0

$$F_1 = F_0 + T_1$$

where

$$T_1 = \frac{GM_m m}{(d-R)^2} - \frac{GM_m m}{d^2}$$

$$\simeq \frac{2GM_m m}{d^3}R \quad \text{since} \quad \frac{R}{d} \simeq \frac{1}{60} << 1$$

For the particle P_3,

$$F_3 = F_0 + T_3 \quad \text{and} \quad T_3 = -\frac{2GM_m m}{d^3}R$$

For the particles P_2 and P_4 the force of attraction is inclined at a small angle α to the line of centres and each can be resolved into a component F_0 along the line of centres and component T_2 or T_4 normal to the line of centres.

It can be shown that

$$T_2 = \frac{GM_m m}{d^3}R$$

Fig. 2.17(a) shows $\mathbf{T}_1$, $\mathbf{T}_2$, $\mathbf{T}_3$ and $\mathbf{T}_4$ which are tide producing forces. They act to cause the water to flow, redistribute itself and attain an 'equilibrium configuration' exhibiting two bulges at points nearest and furthest away from the moon (Fig. 2.17(b)).

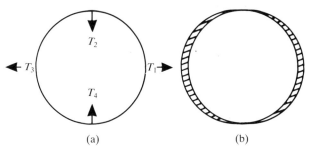

(a)   (b)

Fig. 2.17

2.2(d) Modelling the Earth as a rotating sphere

Very precise measurements of the acceleration due to gravity reveal an interesting phenomenon: g exhibits slight variations at different points on the Earth's surface. In particular a steady increase in g is found with increasing latitude. This seems to contradict Newton's law of gravity, which says that at a point P, a distance R from the centre of the Earth,

$$g = \frac{GM}{R^2}$$

unless, of course, the Earth is ellipsoidal. In fact, the polar radius is indeed about 0.33% smaller than the equatorial radius, but that is not significant enough to account for the measured variations in g. How can this problem be resolved?

The answer lies in the modelling of the Earth. It is necessary to consider the Earth as a sphere, radius $R = 6.378 \times 10^6$ m, rotating with angular speed $\omega = 7.272 \times 10^{-5}$ radians per second about its polar axis.

Consider the point P on the Earth's surface at a latitude λ. If a mass is released from rest just above

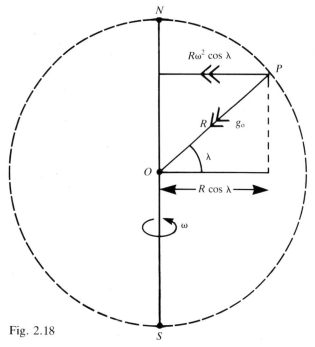

Fig. 2.18

P then its real acceleration due to gravity, relative to the centre of the Earth, O, is g_0 where

$$g_0 = \frac{GM}{R^2} = 9.814 \text{ m s}^{-2},$$

and its direction is vertical.

However measurements of acceleration are taken relative to P which has an acceleration relative to O, of magnitude $R\omega^2 \cos \lambda$ and directed towards the axis ON. The measured acceleration, **g**, relative to P is given by the vector diagram

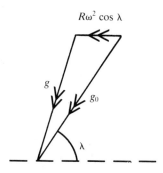

Note that the relative acceleration of the mass is not vertical! Resolving $R\omega^2 \cos \lambda$ along and perpendicular to OP, it is clear that there is a small horizontal component $R\omega^2 \cos \lambda \sin \lambda$. The vertical component of g, say g', is slightly smaller than g_0

$$g' = g_0 - R\omega^2 \cos^2 \lambda.$$

Substituting real values for R and ω gives

$$g' = g_0 - 0.03373 \cos^2 \lambda.$$

When $\lambda = \frac{1}{2}\pi$, $g' = g_0 = 9.814 \text{ m s}^{-2}$; this is the acceleration at the North and South poles.

When $\lambda = 0$, $g' = g_0 - 0.03373 = 9.780 \text{ m s}^{-2}$; this is the acceleration at the equator.

These differences are very small, and for most purposes it is valid to ignore them and simply assume the Earth's surface to be stationary. For very accurate work, however, the refined model incorporating variations in g may be necessary.

Returning to the example of the simple pendulum, $T = 2\pi \sqrt{(l/g)}$ was given as an approximate model for the time period of small oscillations. We can now say that the errors involved when amplitudes are around 15°, errors of approximately 0.4%, are of the same order as errors due to the variation of g at different latitudes.

A further consequence of the modelling of the Earth as a rotating sphere arises with the 'Foucault pendulum', which is just a simple pendulum set up to swing for a long time. In 24 hours the plane of oscillation is observed to rotate through 360° relative to the room in which it is oscillating. Of course, it is the room and the Earth which rotates, the plane of oscillation remains fixed.

2.2(e) Modelling frames of reference as 'inertial frames'

In the above example it could be pointed out that whenever we measure acceleration **relative to the Earth's surface** we make a crucial assumption, that the Earth's surface itself is not accelerating.

The fact that it *is* accelerating slightly implies that Newton's second law is not strictly valid if we measure acceleration relative to the Earth's surface. Newton's laws will only be 'accurate' if we measure acceleration **relative to the centre of the Earth**.

However, it can then be asserted that the Earth itself is accelerating; it rotates about the sun! So, to be even more accurate, we should measure acceleration relative to the sun. This will involve a difference of the order of $d\Omega^2$, where d = radius of the Earth's orbit and Ω is its angular speed; in fact $d\Omega^2 \simeq 0.006 \text{ m s}^{-2}$. Clearly we are dealing with very small inaccuracies now.

Scientists have sought further refinement, because the sun is itself accelerating relative to the rest of the galaxy. And our galaxy is itself accelerating relative to the universe. The question arises, is there a non-accelerating frame of reference out there somewhere? In such a frame of reference, called an 'inertial frame', we could measure all accelerations and know that Newton's laws apply *exactly*.

However, whether this ideal exists is not really important. We use the ideal to model frames of reference with origin at the sun, or the centre of the Earth, or on the surface of the Earth. (This is done in the same spirit as when we use the ideal of a particle to model bodies.) The only important point that must be remembered is that such a model is an approximation and involves errors which may in some situations become significant.

Frame of refence with origin at the	Possible order of errors in accelerations
surface of the Earth	$R\omega^2 = 0.034$ m s^{-2}
centre of the Earth	$d\Omega^2 = 0.0059$ m s^{-2}

- For calculating the motion of a falling apple, the surface of the Earth may be modelled as an inertial frame and assumed not to be accelerating.
- For calculating small variations in g with latitude, the axis of the Earth is assumed not to be accelerating, and axes at the centre of the Earth are modelled as an inertial frame.
- For calculating the orbits of planets about the sun, the sun may be assumed not to be accelerating, and axes at the sun are modelled as an inertial frame.

2.2(f) Non-inertial frames of reference

It has been established that Newton's laws require acceleration to be measured in an inertial frame, that is a non-accelerating frame of reference. This is an idealisation since in practice any frame has an acceleration relative to some other origin. Now if the acceleration of a frame of reference is 'sufficiently small' then it can be legitimately modelled as an inertial frame. But what if the acceleration of a frame is **significant**? Clearly Newton's laws cannot be applied in such a frame of reference and it is referred to as a non-inertial frame.

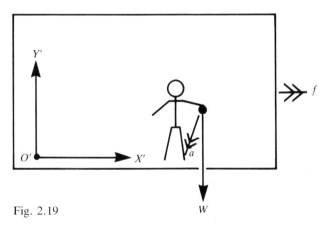

Fig. 2.19

Consider a ball released in an accelerating train. The acceleration of the ball, **a**, **relative to the train** is shown in the diagram. The only force applied is the weight **W**, acting vertically downwards.

Applying Newton's second law in a frame $O'X'Y'$ moving with the train, would give **W** = m**a**, which is clearly not valid.

This problem is resolved by considering acceleration relative to the surface of the Earth (assuming it is an inertial frame), which is **a** + **f**. Then Newton's second law gives

$$\mathbf{W} = m\,(\mathbf{a} + \mathbf{f})$$

which is valid.

The equation can be rearranged in the form

$$\mathbf{W} - m\mathbf{f} = m\mathbf{a}$$

which illustrates that the acceleration in the accelerating frame can be regarded as due to an applied or Newtonian force **W** and a non-Newtonian force, $-m\mathbf{f}$. This is referred to as an inertial force. (See chapter 4, 'Centrifugal force'.)

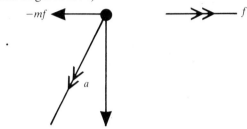

Fig. 2.20

This approach can be applied to any frame of reference with acceleration **f**. Just imagine that all bodies are subject to Newtonian forces *and* an extra inertial force $-m\mathbf{f}$ and from that point on in the analysis, all acceleration can be taken relative to the accelerating frame. This provides a generalisation of Newton's second law to non-inertial frames of reference. If **F** is the resultant applied or Newtonian force and **f** is the acceleration of the frame of reference in which **a** is measured then

$$\mathbf{F} - m\mathbf{f} = m\mathbf{a}$$

In the special case when **f** = 0, or when **f** is negligible compared to other factors in the situation, then this reduces to Newton's second law

$$\mathbf{F} = m\mathbf{a}.$$

2.3 MODELLING CONTACT FORCES

Apart from gravity, other forces of elementary mechanics include friction, normal reaction and tension which arise when two bodies are in contact. At the microscopic level, it is known that electromagnetic interactions arise between atoms in the adjacent surface layers of a body – the interaction between each pair of atoms producing equal and opposite forces by Newton's third law. It is then assumed that the net effect of all such interactions is to produce two equal and opposite forces **S** and $-\mathbf{S}$ as shown in the diagram.

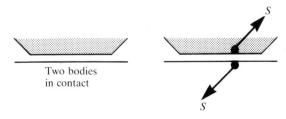

Fig. 2.21 Two bodies in contact with interaction forces **S** and $-\mathbf{S}$.

This resultant contact force **S** may be resolved into normal and tangential components which are referred to as normal reaction and friction respectively.

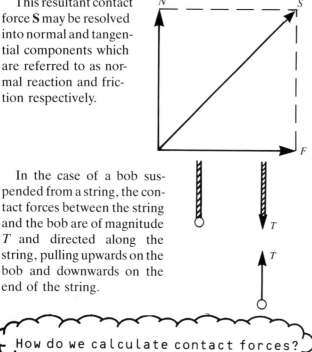

In the case of a bob suspended from a string, the contact forces between the string and the bob are of magnitude T and directed along the string, pulling upwards on the bob and downwards on the end of the string.

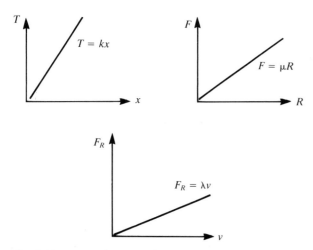

How do we calculate contact forces?

In principle we may analyse the interactions between pairs of atoms, yet any attempt to determine contact forces in this way is a daunting task because of the number and variety of particles involved. We therefore resort directly to experiments in order to measure the contact force in particular situations. For instance experiments show that

- tension (T) in a spring depends on the extension of the spring (x)!
- sliding friction (F) depends on the normal reaction (N)!
- air resistance (F_R) depends on the speed of the body (v) and the viscosity of the air!

A 'model' for the contact force is provided by the mathematical function which best fits the data. Usually the function is linear but only valid over a limited range of parameter values.

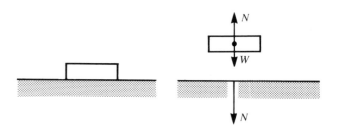

$$T = kx$$

$$F = \mu R$$

$$F_R = \lambda v$$

Fig. 2.22 Linear models for three contact forces.

It is important to appreciate just how limited these models really are. In contrast to the law of gravitation, each model has to be validated for each new practical situation in which we wish to apply them. For this reason they can be described as 'experimental' or 'empirical' models. Clearly they do not have the same range and status as Newton's laws which are central to mechanics.

2.3(a) Modelling friction

As seen earlier, when two bodies are in contact the contact force **S** between them may be resolved into two components: a normal reaction and a tangential (friction) component. In the case of a block resting on a horizontal plane, the contact force is purely nor-

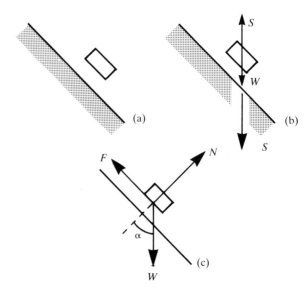

Fig. 2.23

mal in direction. This is confirmed in the force diagram in which only a normal component (**N**) is required to balance the weight and maintain the block in static equilibrium.

If the plane is both rough and sloping and the block does not slide then once again the contact force **S** (exerted on the block by the plane) is vertical but not

Fig. 2.24 A block in contact with an inclined plane (a), showing contact forces (b) and a force diagram for the block (c).

normal to the surface! By resolving **S** into a normal component (**N**) and a tangential component (**F**) static equilibrium is maintained provided

$$N = W \cos \alpha; \qquad F = W \sin \alpha$$

– the magnitudes of both **F** and **N** depend on both the weight W and the angle of inclination of the plane.

There are two cases to consider. The surfaces of contact of the two bodies may be either sliding or relatively static. A model for sliding friction is easier to deal with, though more difficult to validate experimentally!

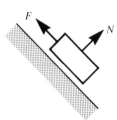

A block at rest on an inclined plane. Surfaces of contact are relatively static.

Fig. 2.25 (a)

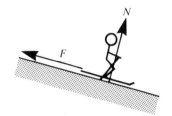

A skier sliding downhill. Surfaces of contact are sliding

Fig. 2.25 (b)

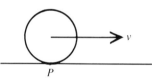

A snooker ball rolling along a table. Surface of contact, (i.e. point of contact) P is static even though centre of the ball is moving

Fig. 2.25 (c)

Sliding friction

1 The friction force is always opposed to the direction of motion of the surface of contact.
2 The magnitude of sliding friction is given by $F = \mu_d N$ where μ_d is called the coefficient of sliding or dynamic friction.

Any model for static friction must take account of there being a range of possible values that friction might take for any normal reaction N. Consider the forces exerted on a block of weight W which is at rest on a rough, horizontal table subject to the push of your hand, **P**. Experimentally it is observed that P can take a range of values, from 0 up to some maximum or limiting value P_L at which the block starts to slide.

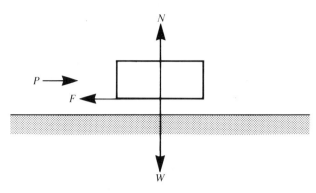

Fig. 2.26 The push **P** on a static block giving rise to a friction force **F**.

But Newton's first law tells us that $F = P$. Therefore F can take a range of values up to a limiting value, 'the limiting value of friction', F_L. In fact **P** can be a pull, also, so that F can be as low as $-F_L$; $-F_L \leq F \leq F_L$. Experiment (see 'The law of friction', chapter 7) suggests F_L is proportional to R over a range of values of R;

$$F_L = \mu N$$

where μ is called the coefficient of static friction.

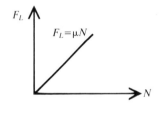

Fig. 2.27 Limiting friction varying linearly with N.

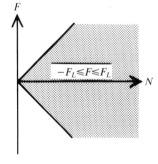

Fig. 2.28

Conclusion

Over a range of N, $0 \leq N \leq N_0$, a model for static friction is given by

$$F \leq \mu N$$

where μ is the coefficient of static friction.

Experiments often show that

■ the graph of F_L against N curves for large N.
■ μ depends on direction. For example, a tyre has different coefficients of sideways slip and forward slip!

Modelling pulleys and pegs

Generally when a string winds around a peg there will be a normal reaction and a friction force. The effect of such a frictional force is to cause the tension in the string to be different on either side of the peg. A peg can actually make an excellent friction

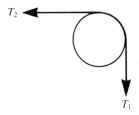

Fig. 2.29

break: wind a rope tightly round a fixed peg or tree and take hold of one end. Invite a friend to pull as hard as they can on the other end. It can be shown that if T_2 is the tension in your part of the rope then, in order to disturb equilibrium, your friend needs to exert a force of magnitude $T_1 = T_2 \, e^{2\pi n \mu}$, where n is the number of windings and μ is the coefficient of friction.

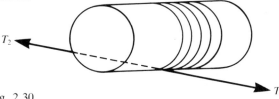

Fig. 2.30

Therefore if the peg is slippery, $\mu \to 0$ and $T_1 \to T_2$, so $(T_1 - T_2)$ is small relative to the other forces involved, i.e. T_1 and T_2.

Conclusion

In this case, we model the peg as 'smooth' and assume negligible difference in tension, $T_1 = T_2$.

The pulley is slightly different, since it can rotate about its bearing with the string. The contact between the pulley and string must be rough or else the string would slide around the pulley! Consider the forces acting on the pulley, which is of mass m, say. Tensions T_2 and T_1 from the string act upon the pulley, tending to cause it to rotate. There is a 'resisting' torque G (due to friction in the bearing) opposing the tendency to rotate. If I is the moment of

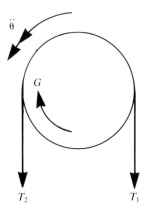

Fig. 2.31

inertia of the pulley about an axis through its centre ($I = \frac{1}{2}ma^2$ for a cylinder of radius a) then the principle of angular momentum yields

$$(T_2 - T_1)a - G = I\ddot{\theta}$$

$$= \tfrac{1}{2}ma^2\ddot{\theta}$$

where $\ddot{\theta}$ is the angular acceleration of the pulley when it has turned through an angle θ.

If the pulley is to remain in equilibrium, $\ddot{\theta} = 0$, then

$$(T_2 - T_1)a = G$$

which means the difference in tensions must be equal to the resisting torque divided by a. Presumably, as with friction, there is a limit to the resisting torque, G, so that rotation eventually occurs as $(T_2 - T_1)$ is increased. If the bearing is 'smooth' this means that G is always small when compared with $(T_2 - T_1)a$. Furthermore, if the pulley is 'light', this implies that m is small and in effect we are assuming that G and $\frac{1}{2}ma^2\ddot{\theta}$ are both small compared with $(T_2 - T_1)a$ and so deduce that

$$T_2 = T_1.$$

Conclusion

A suitable model for a pulley may be that it has a 'smooth' bearing and is 'light', in which case the tension in the string is the same on both sides of the pulley.

2.3(b) Modelling tension in a string

A mass hangs in equilibrium on the end of a string connected to a ceiling.

    What forces act?
    What is meant by 'tension in the string'?
    What is meant by 'breaking tension'?

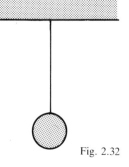

Fig. 2.32

We can perform several force analyses. Assume the string has weight $w = mg$, the mass has weight $W = Mg$; let the force of the ceiling on the string be $\mathbf{F}_1$ and the force of the string on the mass be $\mathbf{F}_0$ as shown. Consider the forces on the combined string–mass (see Fig. 2.33(a)) and apply Newton's first law.

$$F_1 = w + W,$$

the tension at the top of the string is simply the com-

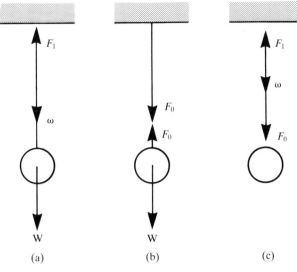

Fig. 2.33

(a)   (b)   (c)

bined weight of the string and the mass. Similarly, from Fig. 2.33(b)

$$F_0 = W,$$

the tension in the bottom of the string is equal to the weight of the mass. From Newton's first law for the string, Fig. 2.33(c) gives

$$F_1 = w + F_0.$$

Now consider what happens if W is very large, so that the string breaks. Clearly the string becomes **two bodies**. Consider the string as two bodies even before it breaks; what holds the two parts of the string together? The answer is more obvious for a chain than a string, but the string can just be thought of as a chain of strands rather than links.

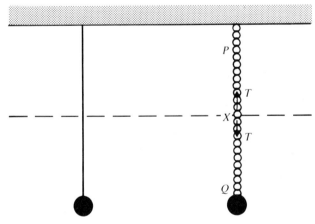

Fig. 2.34

Conclusion

There are internal forces (tensions) within a string or chain which hold the string or chain together. There is a limit to the tension which the links of a chain or strands of a string can sustain before it will break.

We can be precise about this and define the tension T at a point, X, of the string as being the force of interaction between PX and XQ. Suppose the length of string is l, PX is of length x, XQ is of length $l - x$. Let the density of the string be ρ per unit length, so $\rho l g = w$. Then the weight of XQ is $\rho(l - x)g$, and $T = W + \rho g(l - x)$.

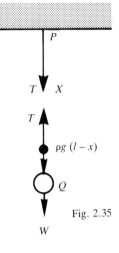

Fig. 2.35

Therefore

$$T(0) = W + w$$

and

$$T(l) = W.$$

Essentially, the tension in the string increases from W at the bottom to $W + w$ at the top. The graph is linear if the density is uniform.

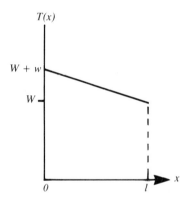

Fig. 2.36   Variation of tension $T(x)$
along the length of the string

Conclusion

In any situation where a force is applied to the lower end of the string we can expect the tension to increase linearly up to the top end of the string

$$T(x) = W + \rho g(l-x).$$

$$\text{for } 0 \leq x \leq l$$

33

Modelling a 'light' string

When the weight of the string is relatively small compared with the other forces involved, we say 'the string is light' and proceed to assume its density to be zero!

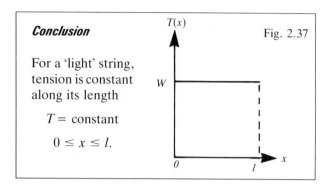

Conclusion

For a 'light' string, tension is constant along its length

T = constant

$0 \leq x \leq l$.

Fig. 2.37

Note that the **absolute** weight of the string is not in question! We are concerned only that the weight is **relatively** small, say only a small fraction of the weights of other bodies in the analysis.

In practice it can be demonstrated that the *same* string may reasonably be assumed 'light' in some situations but not in others! Consider the cable of a tug. When the forces applied are small, the cable falls under its own weight in a curve which is called a catenary. When the tug starts pulling, the cable takes up virtually a straight line, as if it were weightless.

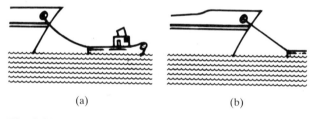

(a)   (b)

Fig. 2.38

In Fig. 2.38(a) the weight of the cable is significant, while in Fig. 2.38(b) the weight of the cable is negligible.

This illustrates a basic principle of modelling: ignore masses and forces which are small **relative** to the main determining masses and forces.

Modelling tension in an elastic string or spring

Normally when a string or spring is under tension, it will stretch.

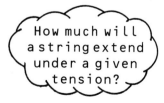

How much will a string extend under a given tension?

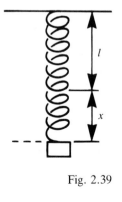

Fig. 2.39

For a given spring or string we can apply weights and plot extension as a function of tension (see 'Hooke's law', chapter 10). Hooke found a linear relationship, $T = \kappa x$ over a limited range of values where κ is the spring 'stiffness'. In reality, most manufactured springs have some 'pre-tensioning', and

$$T = T_0 + \kappa x.$$

Very few springs have a 'Hookean' relationship when under compression as well as under extension.

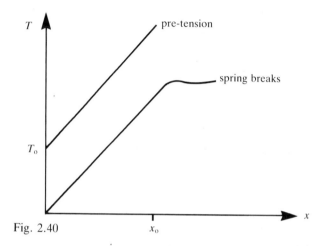

Fig. 2.40

All springs and strings will break or distort when the force applied becomes too great. Nevertheless, a good modelling assumption for many springs and strings is:

For a limited range $0 \leq x \leq x_0$, $T = \kappa x$.

There is a problem with Hooke's law. If a weight mg produces an extension a in a string of natural length L, then the same weight produces an extension $2a$ in a string of length $2L$. In each case the tension is the same yet the string constant κ differs, so κ is length dependent. This situation is remedied by defining $\kappa = \lambda L$ where λ is called Young's modulus of elasticity. (See 'Elasticity', chapter 10.)

$$T = \frac{\lambda x}{L}$$

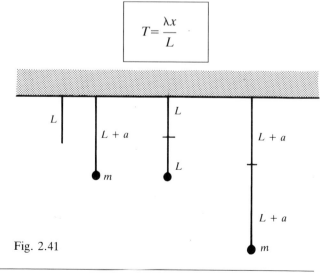

Fig. 2.41

What is
an inextensible string?

All materials have some elasticity (even a metal bar will extend under some stretching force!). In many cases, however, a string will extend by only a negligible amount under the forces involved. It often makes sense then to ignore the extension and in such cases we model the string as 'inextensible'. If $x_0 \ll L$, or if κ is so large that x remains small in a given problem, then we model a string as being **inextensible**.

2.3(c) Modelling air resistance

How large is the force
due to air resistance?

Any cyclist knows from experience that air resistance

■ tends to slow the bike down.
■ has little effect at low speeds but becomes more noticeable at higher speeds.
■ can be reduced by crouching over the handlebars.

It follows that air resistance is a force whose magnitude F_R depends on a body's speed, shape and size. In fluid mechanics the Stokes' drag experienced by spheres positioned in a stream of viscous fluid, moving with uniform speed U, has been found to be proportional to $(\mu D U)$ where μ is the viscosity of the fluid and D the diameter of the sphere. This relationship assumes that inertia effects are negligible, that is, that the speed of the fluid is 'relatively small'. The precise nature of this dependence of F_R on speed must be settled by experiment. Indeed experiments have been performed on smooth spheres, positioned in a wind tunnel with air blown past them with speed v, and the air resistance measured.

Data is available over a large range, $10^{-10} < F_R < 1$ newton, and the diagram shows a plot of $\log_{10}(F_R)$ against $\log_{10}(Dv)$ which indicates that F_R does increase with Dv. Although there is no simple formula relating F_R, D and v the dotted lines do suggest approximations which can be used over limited ranges.

Linear approximation: for $Dv \leq 10^{-5}$

$$\log_{10}F_R = -3.77 + \log_{10}Dv$$

$$\boxed{F_R = k_1 Dv; \quad k_1 = 1.7 \times 10^{-4}}$$

Quadratic approximation: for $10^{-2} < Dv < 1$

$$\log_{10}F_R = -0.7 + 2\log_{10}Dv$$

$$\boxed{F_R = k_2(Dv)^2; \quad k_2 = 0.2}$$

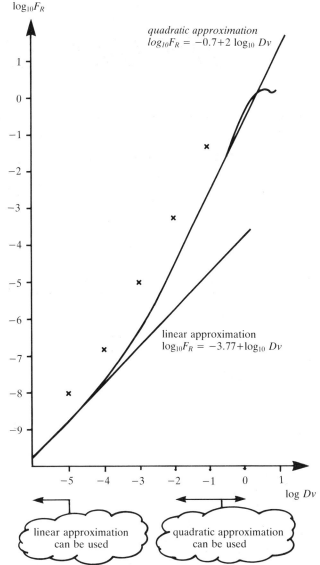

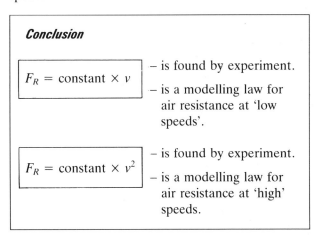

Fig. 2.42

For bodies other than spheres, for example cylinders, discs, ellipsoids, experiments suggest similar linear (quadratic) dependence of F_R on v at low (high) speed.

Conclusion

$F_R = $ constant $\times v$	– is found by experiment.
	– is a modelling law for air resistance at 'low speeds'.
$F_R = $ constant $\times v^2$	– is found by experiment.
	– is a modelling law for air resistance at 'high' speeds.

3
THE MODELLING PROCESS

The modelling process is a systematic approach to problem solving which begins by specifying a problem!

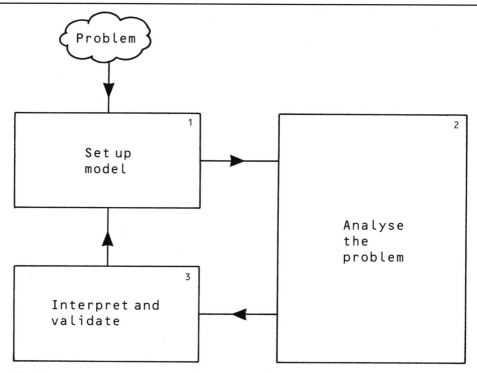

Fig. 3.1

---

aims of chapter three

After reading this chapter you will:

- be aware of the modelling process as a systematic approach to problem solving.

- appreciate the importance of a **real problem** for motivating study; in particular for initiating an investigation based on modelling.

- appreciate the three stages of the modelling process in Newtonian mechanics.

- have three examples, with full documentation, of the modelling process applied to real problems.

3.1 A 3-STAGE MODELLING DIAGRAM

In chapter 2 we saw how modelling is at the heart of the subject of mechanics; that the solution of real problems requires us to use models for bodies and for the forces which act on them. Common assumptions are:

- the body is a particle,
- gravity is constant,
- the string is light and inextensible,
- the string is extensible and the tension obeys Hooke's law,
- the pulley is light and smooth,
- air resistance is negligible.
- friction is negligible,
- friction obeys the law $F \le \mu R$.

In any given problem we have to **select** appropriate assumptions like these in order to retain the essentials of the situation and ignore the inessentials. We then apply Newton's laws and solve the problem using pure mathematics, that is technique. Finally, we need to interpret the meaning of the mathematical solution and check that it predicts what is observed in the real world. Solving real problems in this way is called 'modelling'. It usually involves three distinct stages which together constitute a 'modelling process' as represented by Fig. 3.1.

We will now look at each stage in turn, before considering some examples of the complete process. But first, the starting point must be a problem and usually a 'real problem'!

By a real problem we mean one which has been identified in the real world and is not yet transformed into a mathematical problem. For instance

- How high can I throw this golf ball, and how long will it take to fall back to Earth?
- What is the time period of the oscillations of a simple pendulum?
- How high above the Earth's surface is the orbit of a geostatic satellite? (A satellite which is static relative to the point directly beneath it on the Earth's surface.)

Of course, there are plenty of 'problems' which do not require modelling, such as

- How many factors has p^n?
- Differentiate $y = |x|$.
- Given $x(0) = 0$ and $x(0) = a$, integrate $\ddot{x} = -\omega^2 x$.

These problems are already expressed in **mathematical** terms. It is not always clear that they relate to a 'real world' problem, and if they do, what that problem is.

The first stage in solving a real world problem must be to set up a model so that it can take a mathematical form.

Set up model	1

As already stated, the aim is to set up a model which simplifies the real situation while retaining its important features – and this is achieved by making various assumptions; for example

- assume the body is a particle.
- assume gravity is constant.
- assume string is light.
- assume the collision is perfectly elastic.

As part of building the model we also draw a diagram, label the forces acting and define the variables in the problem. To set up a model, therefore, we must:

(i)   specify the assumptions.
(ii) draw a diagram; label the forces.
(iii) introduce appropriate variables.

Analyse the problem	2

Here the aim is to analyse the problem: that is, to formulate a problem in mathematical terms and seek a mathematical solution.

In Newtonian mechanics the fundamental assumption is that Newton's laws (of motion and gravitation) are valid. Consequently a mathematical problem emerges which often involves one or more equations to which a solution is sought. The analysis then involves solving these equations using techniques such as

- eliminating unknowns from simultaneous equations,
- integrating differential equations.

Having found a solution to the mathematical problem the next step is to interpret that solution by identifying and exploring its consequences in real terms.

Finally, there is a need to test the validity of this solution to the mathematical problem. If it does compare well with reality then clearly the initial assumptions made (in simplifying the real situation and setting up a model) were justified. On the other hand if 'significant differences' do arise then one or another of the assumptions made needs to be re-examined and the model refined. In this context note the presence of an arrow connecting **stage 3** to **stage 1** in the modelling diagram (see Fig 3.1).

3.2 REAL WORLD PROBLEM SOLVING VIA MODELLING

In this section we shall illustrate the modelling process by its application to three real world problems.

3.2(a) Dan KoKo's record dive into an air bag

The *Guiness Book of Records* records that

'the greatest height reported for a dive into an air bag is 326ft = 99.36m by Dan KoKo from the top of the Vegas World Hotel, on 13th August, 1984'.

Problems of interest are: what was the time of flight? what was his speed on impact?

Assume that:
(i)   Dan Koko is a particle of mass m.
(ii) his weight is constant, mg.
(iii) air resistance is negligible.
(iv) his initial speed $U=0$, time of flight is T, impact speed is V.
(v)   the motion takes place in a vertical line.

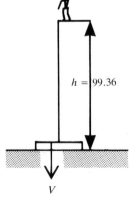

Note that none of these assumptions is a statement of fact; they all involve an approximation to reality. This involves a simplification in which certain factors (air resistance, variable gravity, take off speed, Dan's height) are assumed to be negligible.

Given this model it is simple to apply kinematical equations like (3.3) and (3.4) below. But for the purposes of later generalisation, we shall apply Newton's second law:

$$mg = m\,\frac{dv}{dt} \qquad (3.1)$$

or

$$mg = mv\,\frac{dv}{dy} \qquad (3.2)$$

subject to initial conditions $t=0$, $y=0$, $v=0$.

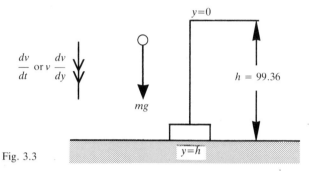

Fig. 3.3

Integration of (3.1) gives $mgt = mv+c$ where $c = 0$ from the initial conditions.
Therefore

$$v = gt$$

and when $t = T$, impact speed

$$V = gT. \qquad (3.3)$$

Integration of (3.2) gives $mgy = \frac{1}{2}mv^2 + c_1$ where $c_1 = 0$ from the initial conditions.
Therefore

$$v^2 = 2gy.$$

Clearly when $t = T$, $y = h$ and $v = V$

$$V^2 = 2gh. \qquad (3.4)$$

If $g = 9.81$ m s^{-2} then $V = 44.1$ m s^{-1}

and $T = 4.50$ s

1 The solution gives an impact speed of 44.1 m s^{-1} (or 159 km/hour) and time of flight 4.5 seconds. Records of the stunt give the actual impact speed as 141 km/hour – so our result is within 13%!
   Equation (3.4) can be obtained by applying 'conservation of mechanical energy': $\frac{1}{2}mV^2 = mgh$ – so the 13% error may be due to energy losses.
2 The result is mass-independent and $V \propto \sqrt{h}$ implies that small changes in h lead to even smaller changes in V.

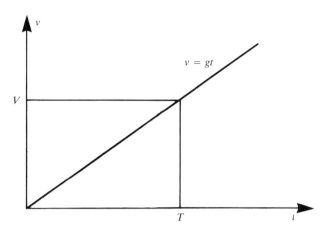

Fig. 3.4

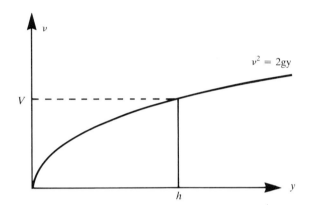

Fig. 3.5

3 The result depends on g. If the stunt were to be conducted on the moon, a reduced g would give a reduced V but an increased T!

4 A refined model would include air resistance to account for the energy loss mentioned above. A resistance force $R = kv$ as suggested in chapter 2, section 2. 3(c) means that (3.1) and (3.2) become

$$mg - kv = m\frac{dv}{dt}$$

$$mg - kv = mv\frac{dv}{dy}$$

This refined model could now provide the starting point for a second round of analysis, interpretation and validation.

3.2(b) Finding the period of a pendulum

Our second illustration of the modelling process is one that fascinated Galileo.

'At the age of 17 Galileo was sent to the University to study medicine. One day, while attending a service in the cathedral at Pisa his mind was distracted by the great bronze lamp suspended from the high ceiling. The lamp had been drawn aside in order to light it more easily and when released it oscillated to and fro

with gradually decreasing amplitude. Using the beat of his pulse to keep time, he was surprised to find that the period. . .was independent of the size. . .of the oscillation. Later, he showed that the period. . . depends solely on the length of the pendulum. . .the result was that. . .(Galileo) devoted himself to science and mathematics' (Eves, (5)).

Problem How does the time period of a pendulum depend on its length?

Set up model	1

As discussed earlier the principal assumptions are:

(i)   the string is light and inextensible of length l.

(ii) air resistance is negligible.

(iii) the mass attached is a particle of mass m.

(iv) the amplitude of the oscillations ($\theta = \theta_0$) is small such that $\sin\theta \simeq \theta$ for all θ, where $\theta(t)$ is the angle the string makes with the downward vertical at time t.

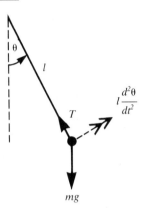

Fig. 3.6

In the tangential direction the mass has speed

$$v = l\frac{d\theta}{dt}$$

and acceleration

$$a = l\frac{d^2\theta}{dt^2},$$

Analyse the problem	2

Apply Newton's second law in the tangential direction:

$$-mg\sin\theta = ml\frac{d^2\theta}{dt^2}. \qquad (3.5)$$

The mathematical problem is to solve equation (3.5) subject to initial conditions $\theta = \theta_0$ and $\frac{d\theta}{dt} = 0$ at $t = 0$.

Equation (3.5) gives

$$\frac{d^2\theta}{dt^2} = -\frac{g}{l}\sin\theta.$$

Assumption (iv) that θ_0 is small, so that $\sin \theta$ is approximately θ, reduces the above equation to the simple harmonic motion equation,

$$\frac{d^2\theta}{dt^2} + \omega^2\theta = 0, \text{ where } \omega = \sqrt{(g/l)}.$$

The appropriate solution satisfying the initial conditions is $\theta = \theta_0 \cos \omega t$ and the time period is

$$T = \frac{2\pi}{\omega} = 2\pi \sqrt{\left(\frac{l}{g}\right)}$$

Interpret and validate [3]

1 The time period T is independent of mass m, and varies with the square root of l. Therefore, as l is doubled, T is increased by a factor of $\sqrt{2}$. This can be easily validated in practice. (See 'The simple pendulum', chapter 10.)

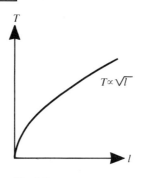

Fig. 3.7

Also we can vary m and find no significant variation in T in practice, as long as the mass is **large enough to keep the string taut** but not so large that the string stretches or breaks.

2 g can be calculated by plotting T against $\sqrt{l}$ to obtain the gradient. Historically, g was calculated by this means by Huygens using pendulums, and was found to vary at different points on the Earth's surface.

3 Using $g \simeq 10$, we can deduce that $l = \frac{1}{4}$m gives a time period of 1 second. This works well in practice, but care must be taken to measure l from the **centre of mass** of the bob which may have a non-negligible diameter. (See 'The simple pendulum', chapter 10.)

4 The solution $\theta = \theta_0 \cos \omega t$ does not predict oscillations which decay. An improved model, including air resistance, is therefore needed. See 'Amplitude decay', chapter 10, where it is found that the time period remains constant even though the amplitude decays.

5 In chapter 2, section 2.1 it was stated that our model holds good for amplitudes up to approximately 15°; in fact the percentage increase in T at 15° is only 0.4%. For larger amplitudes the solution to the full pendulum equation

$$\frac{d^2\theta}{dt^2} + \frac{g}{l} \sin \theta = 0$$

is required. This constitutes a refined model which can lead to a second round of analysis, interpretation and validation.

3.2(c) Positioning a communications satellite

As a final example, we consider a problem of modern interest, the positioning of communications satellites. Ideally, these are placed so that they orbit the Earth once a day above the Earth's equator. This means that a satellite remains static relative to the point P directly below it on the Earth. For this reason they are called geostatic.

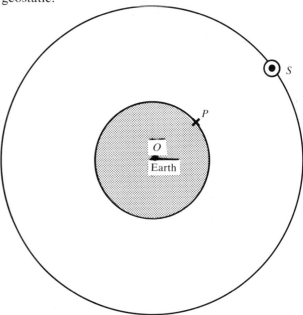

Fig. 3.8

Geostatic satellites orbit the Earth (centre O) so that the points O, P, S remain collinear.

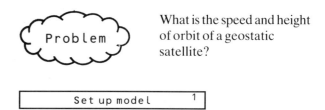

What is the speed and height of orbit of a geostatic satellite?

Set up model [1]

Assume that:

(i)   the Earth and satellite are particles of mass M and m.
(ii) the Earth is stationary yet rotates about its axis once per day; the satellite rotates about the Earth in a circular orbit of radius d with angular speed ω.
(iii) The gravitational force has magnitude $F = GMm/d^2$ where $G = 6.673 \times 10^{-11} \text{ m}^3/\text{kg}^{-1}\text{s}^{-2}$ and $M = 5.97 \times 10^{24}$ kg. We wish to find d so that the time of orbit $T = 1$ day $= 86\,400$ s, where $T = 2\pi/\omega$; so $\omega = 2\pi/T = 7.27 \times 10^{-5}$ s^{-1}.

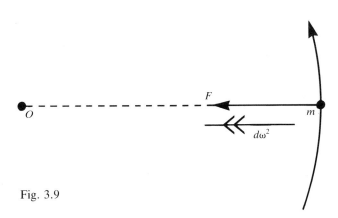

Fig. 3.9

    Analyse the problem   2

Newton's second law applied to the satellite gives

$$F = \frac{GMm}{d^2} = md\omega^2$$

$$\Rightarrow \quad [GM = d^3\omega^2]$$

$$\Rightarrow \quad d^3 = \frac{GM}{\omega^2} \quad \text{or} \quad d = \left(\frac{GM}{\omega^2}\right)^{\frac{1}{3}}.$$

In terms of T,

$$d^3 = \frac{GMT^2}{4\pi^2}.$$

For $T = 86400$ s, $\omega = 7.27 \times 10^{-5}$

$d = 42.3 \times 10^6$ m to 3 s.f.

    Interpret and validate   3

1 Since we modelled the Earth as a particle, we must subtract the radius of the Earth from the radius of orbit d to obtain the satellite height – 35 900 km. This is in fact the height of geostatic satellites, the first of which, called Syncom II, was launched in 1963.

2 The formula for d is independent of mass: *all* geostatic satellites must have the same orbit height irrespective of mass. Even if the satellite breaks up, we might expect the debris to orbit at the same height.

3 We can use this model to obtain any orbit heights. For instance, a satellite with period 1 month = 27.3 days gives $d = 380\ 000$ km, which will be recognised as the distance of the moon from the Earth!

4 The formula $d^3 = \dfrac{GM}{\omega^2} = \dfrac{GM}{4\pi^2}T^2$ applies to any orbit problem, where M is the mass at the centre of the orbit. Consequently, it can be applied to planets orbiting the sun, with $M_s = 1.99 \times 10^{30}$ kg. For $T = 1$ year $= 3.16 \times 10^7$ seconds, one obtains $d = 1.5 \times 10^{11}$ m, the Earth's mean orbital distance from the sun.

5 The formula $d^3 \propto T^2$ is Kepler's law for all elliptical orbits around the sun where d is the length of the semi-major axis. We have proved this for circular orbits; Newton proved it for elliptical orbits in the *Principia*. See Chapter 1, Section 4.

   A graph of d against T for the data in the table agrees well with our model (see fig. 3.10).

	$d(\times 10^6$ km)	Period ($\times 10^6$ seconds)
Mercury	57.9	7.62
Venus	108	19.4
Earth	150	31.6
Mars	228	59.4
Jupiter	778	376
Saturn	1430	932
Uranus	2870	2654
Neptune	4500	5214
Pluto	5910	7839

$$d = (GM_sT^2/4\pi^2)^{\frac{1}{3}}; \quad GM_s/4\pi^2 = 3.36 \times 10^8.$$

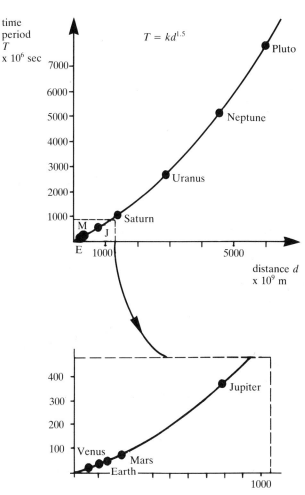

Fig. 3.10

3.3 CONCLUSIONS

The 3-stage modelling diagram is essentially a convenient simplification of a more general 7-part modelling diagram consisting of 4 'locations' and 3 'stages' or processes.

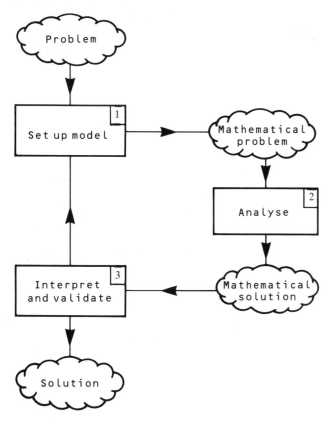

When solving a real problem which requires modelling, the procedure is the following:

(a) **First stage** 'Setting up the model' is the process which takes one from the real problem to a mathematical problem.

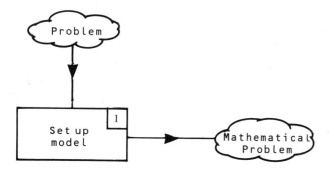

(b) **Second stage** 'Analyse the problem' is the mathematical part of problem solving. It is an analytical process by which we proceed from a mathematical problem towards a mathematical solution.

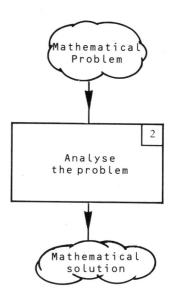

(c) **Third stage** 'Interpret and validate' is the final process in which the mathematical solution is interpreted and validated so as to produce a **solution** to the original problem. Do note, however, that after validating and comparing with reality the solution may be either inadequate or insufficiently accurate. It will then be necessary to refine the model by returning to stage 1 and going around the modelling circuit once more!

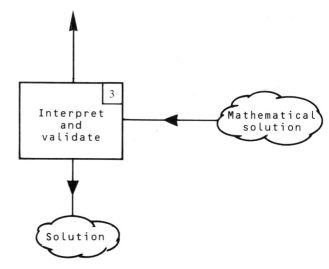

The 3-stage modelling diagram is an attempt to represent this whole modelling activity in a simple way which emphasises the three distinct processes.

- setting up a model

- analysing the problem

- interpreting and validating the solution.

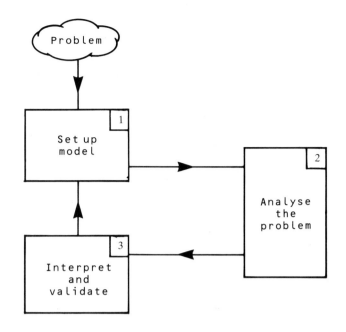

Too often in traditional mechanics courses the emphasis has been firmly focused on deriving solutions to well defined mathematical problems, which means the activity has been centred on stage 2 of the modelling diagram.

Besides the ability to apply mathematics, modelling requires the applied mathematician to acquire additional, complementary skills, namely,

(a) setting up a model – making assumptions and keeping them 'in view'.
(b) interpreting solutions/validating solutions – understanding what the solution means and what are its consequences, and how it compares with reality!

Whenever problems arise in *real* situations these skills are essential. The simple practical problems involving apparatus outlined in chapters 5 onwards require modelling even though sometimes the model in use is very simple. The solutions provided are, therefore, written up with the 3-stage diagram in mind, and with the emphasis on stages 1 and 3. In many cases attention must be given to explaining significant differences between practical results and the theoretical model. (Very few models are quite as accurate as Newton's model for gravity and planetary motion!) In such cases refinements to the models can often improve the situation. Such refinements usually involve introducing more complex mathematics and technique. Introducing air resistance, for instance, in our first example, motivates the study of differential equations such as

$$-kv - mg = \frac{mdv}{dt}$$

We hope that a modelling approach will, therefore, not only develop a broader view of mathematics and a set of problem solving skills, but also excite interest and curiosity in further mathematical study.

4
MISCONCEPTIONS

An English family visiting the Sudan took their young son each day to see the statue of General Gordon in Khartoum. On their last day, the young lad went with his mother to say farewell to General Gordon. Coming away he looked puzzled and his mother asked what was wrong.

'Who is that man', he said, 'sat on General Gordon?'

aims of chapter four

At the end of this chapter you will:

■ know that misconceptions in mechanics are widespread, inevitable and in need of continued attention throughout the A-level course (and beyond!).

■ be able to recognise some of the more common and change-resistant misconceptions:

 (a) Aristotelian misconceptions of force and motion.
 (b) misconceptions of centripetal and centrifugal force in circular motion.
 (c) misconceptions of the vector qualities of displacement, velocity and acceleration.
 (d) misconceptions about Newton's third law and its application to bodies in contact.
 (e) misconceptions about friction such as 'friction always opposes motion'.
 (f) misconceptions about 'weight' and 'weightlessness'.

■ have a selection of diagnostic questions which can be used with students and colleagues to reveal misconceptions and stimulate discussion.

contents

The story of General Gordon, an earlier version of which appears in References, Shah (6), serves to illustrate how easy it is to misunderstand and 'get the wrong idea'! A moment's reflection will reveal that misconceptions are ubiquitous. They feature in the learning process and are encountered by students of every discipline, whether it be mathematics, music or mechanical engineering.

Well known examples include:

- the notion that the Earth is flat!
- the Earth is centre of the universe with the sun and planets orbiting around!
- heavy objects 'fall faster' than light ones!
- high tide at Dover implies low tide at Calais!

4.1 MISCONCEPTIONS IN MECHANICS

Misconceptions in mechanics are often deep-rooted and for many pupils change-resistant despite many years of tuition in the principles of mechanics and their application. At the University of Leeds, data was collected over a ten year period, 1978–1988, by means of a diagnostic questionnaire given to first year undergraduates in mathematics and engineering. This clearly showed that misconceptions and confusion are commonplace – even amongst students attaining high grades at A-level. For example, students often have quite erroneous ideas about the nature of force and motion and even in the most simple situations they often draw forces incorrectly on a diagram. There is a real danger that unless such misconceptions are given serious attention and time, students will not be able to tackle problems outside a very small repertoire of stereotypes which they have seen many times before.

Many of the misconceptions are deeply rooted in our subjective experience of the physical world, particularly (a) and (b) as listed in the aims of this chapter. Some are rooted in language and the conflict between the scientific and everyday usage of terms like velocity, speed, acceleration and weight. Other misconceptions such as (d) and (e) and to some extent (c) are a result of inappropriate **teaching**.

In this chapter a sad face 😦 indicates a misconception and a happy face 😀 indicates a correct answer.

Implications for effective teaching

The effective teaching of Newtonian mechanics naturally requires that a student's misconceptions and sources of confusion be recognised and transcended as an integral part of the learning process. We believe than an optimum teaching strategy for achieving this aim involves:

(i) being prepared, as teachers, to listen to students so that their misconceptions may emerge and be identified.

(ii) providing situations in which students are encouraged to formulate their ideas. This can be done using questions such as those in the questionnaire but **even better** by presenting a practical, mechanical situation for small groups or the class to investigate and analyse. The practical investigations in chapters 5–11 will provide many such examples.

(iii) encouraging discussion and debate between students and between students and teacher.

Identifying misconceptions

For the purpose of identifying misconceptions, the mechanics questionnaire on the following four pages provides a set of 17 illustrative examples, drawn from various sources, which have proved to be useful. Sections 4.2–4.6 contain an analysis of some of the relevant misconceptions and sources of confusion.

1

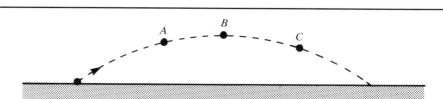

A ball is projected under gravity as shown. If air resistance is negligible insert an arrow at each position A, B, and C to indicate the direction of the resultant force acting.

2

A ball is projected vertically upwards. It rises through position D until it reaches its highest point E and then falls back down through F. Mark on the diagram an arrow which shows the direction of the resultant force acting at each position.

Neglect air resistance.

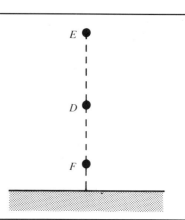

3

A bob on the end of an inextensible string swings in an arc between R and S. Ignoring air resistance insert the direction of the resultant force when it is at the lowest point T.

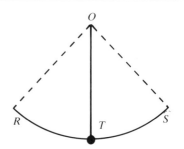

4

A bob, of mass m, is attached to a light and inextensible string and rotates in a horizontal circle of radius r with an angular speed ω about the vertical. Ignoring air resistance insert the forces acting on the bob.

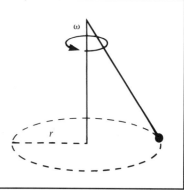

5

A fly, of mass m, is at rest relative to a turntable which rotates with constant angular speed ω.
    Insert the forces acting on the fly.

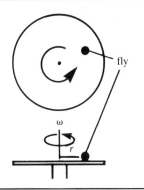

6

A marble, of mass m, can move freely along a smooth groove in a turntable which rotates with constant angular speed ω. Insert the forces acting on the marble.

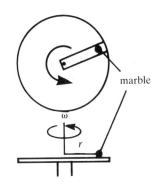

7

(a) 'If speed is constant, then acceleration is zero!'

(b) 'Average velocity cannot be zero over each rotation if its average speed is $a\omega$.'

    True or false?   Comment.

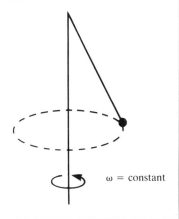

8

A car travels with steady speed along a winding road.
Is its acceleration: zero?
                      constant?
                      variable?
If the car now travels along this winding road with increasing speed, insert the direction of its acceleration at A, B and C.

9

(a) 'You have put $\ddot{x}$ downwards: surely this is incorrect since it is really accelerating upwards.'

(b) 'How can a length x be *negative*?'

(c) 'The least speed is zero, which occurs when $\ddot{x} = 0$ according to calculus . . . '

Comment!

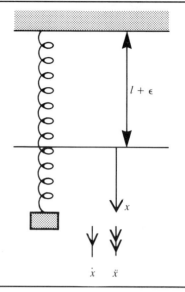

10

A light and inextensible string is attached to the block as shown. Weights are gradually added to the holder and the block of mass m remains at rest subject to the forces shown.

What is the relationship between F and R?

Insert the force(s) exerted by the block on the table.

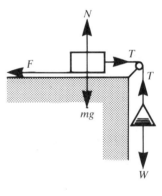

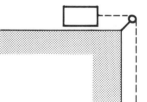

11

A car travels at constant speed along a straight and horizontal road. If the car experiences a force due to air resistance R, what other force must act on the car?

Insert the force on the diagram.

12

A car moves in a circle at constant speed subject to air resistance of magnitude R. Indicate by an arrow the direction of the friction force exerted by the road.

13

A snooker ball rolls down an inclined plane and experiences weight **W**, normal reaction **N** and a frictional force, **F**.

What can you say about its total energy: kinetic and potential?

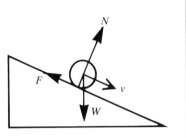

14

What force do bathroom scales measure?

15

Do the scales read heavier or lighter when you are in a lift accelerating upwards?

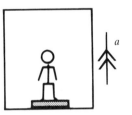

16

A woman is in free fall holding her suitcase. Why does it feel weightless when air resistance is negligible?

17

An astronaut orbiting the Earth in a spaceship feels weightless. Why?

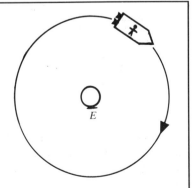

4.2 FORCE

4.2(a) Aristotelian misconceptions

Before Newton, most scholars subscribed to the teaching of Aristotle (384–322 BC) on force and motion, which might be summarised in the two statements:

■ the natural state of matter is that of rest.
■ if a body moves there must be a force.

For Aristotle, forces caused bodies to move and their absence implied that bodies remained at rest. Such ideas are indeed consistent with parts of our experience – the pushes and pulls with which we are familiar.

Push a box and it moves.   Stop pushing and it stops moving.

It is perhaps not surprising therefore that many students bring with them to the study of mechanics a set of intuitive ideas about force which are essentially Aristotelian:

■ if there is no force, there is no motion!
■ if there is motion, then there must be a force!
■ force acts in the direction of motion!

Since these statements are true in a number of specific and familiar situations there is a tendency to assume that they are generally true. The student is then in the grip of misconceptions which produce answers such as the following to questions 1–3.

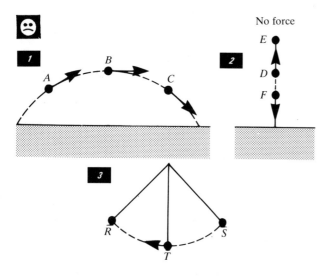

These answers are consistent with Aristotle's ideas and are intuitive and yet they are clearly incorrect. Compare them with the correct answers according to Newton's theory.

Correct answers

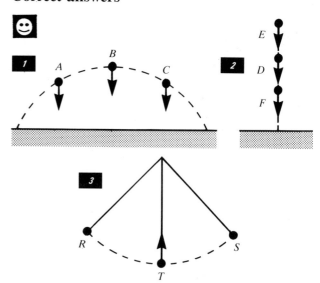

In addition we have the following from Newton's laws of motion.

1 Force causes a change in motion or momentum
Force is proportional to the rate of change of linear momentum. Thus, at D in question 2, the motion is vertically upwards yet the speed is decreasing. So its change in momentum is negative (directed downwards); the mass is accelerating towards the Earth in the direction of the applied force.

2 Force is the action of one body upon another
There is always a body causing a force to be exerted. In questions 1 and 2 this other body is the Earth and the force is due to gravity, i.e. weight.

These two characteristics of force should be explicitly discussed and argued through with classes meeting Newton's laws for the first time. Clearly, understanding force is an essential step towards understanding Newton's laws. Here is a short test of the concepts.

Give an example of a situation where:

(a) the applied force is at right angles to the direction of motion.
(Answer: the ball at B; pendulum at T in questions 1 and 3.)

(b) a body is stationary and yet acted upon by an applied force.
(Answer: the ball at E; pendulum both at R and S in questions 2 and 3.)

(c) a body has constant, non-zero velocity and no resultant force acts upon it.
(Answer: box sliding across the table under the action of a push and friction; rocket in outer space, where gravity forces of planets are negligible.)

4.2(b) Centripetal force : centrifugal force

Fact or fiction? Real or imaginary?

Centripetal force

In question 4 students may insert a 'centripetal force $mr\omega^2$' as in Fig 4.1(a) or a 'centrifugal force, $mr\omega^2$' as in Fig 4.1(b).

 The correct response, in accordance with Newtonian theory, is that the applied forces on the mass are tension **T** and gravity $m\mathbf{g}$;

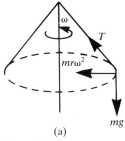 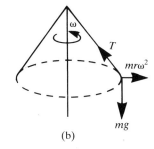

(a)   (b)

Fig. 4.1

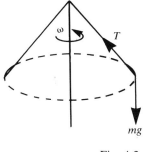

☺ Correct answer

This problem has been given to several groups of sixth form students and it is useful to consider two typical student responses as to why they introduced centripetal or centrifugal forces.

Fig. 4.2

'I insert $mr\omega^2$ in Fig 4.1(a) because it is the centripetal force which causes circular motion. If there is no force towards the centre then why does it move in a circle?'

Newton's second law: $\mathbf{F} = m\mathbf{a}$ means that the resultant applied force is equal to mass × acceleration. Here the resultant applied force is the vector sum of **T** and $m\mathbf{g}$. There is no applied force in addition to **T** and $m\mathbf{g}$.

The left hand side of $\mathbf{F} = m\mathbf{a}$ is therefore the vector sum $(\mathbf{T} + m\mathbf{g})$ which acts towards the centre.

The right hand side is $mr\omega^2$ directed towards the centre which represents the mass × acceleration caused by the applied forces **T** and $m\mathbf{g}$. However it is confusing and therefore, in our opinion, bad practice to place

RESULTANT

Fig. 4.3

both the applied forces and their resultant on the same figure. Students simply need to be aware that the effect of **T** and $m\mathbf{g}$ in Fig. 4.3 is the same as that of the resultant **R** in Fig. 4.4.

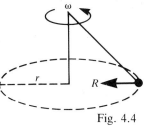

Fig. 4.4

Centrifugal force

'I insert a centrifugal force, $mr\omega^2$ in Fig.4.1(b) because this is the force which causes the mass to go outwards – and stops it falling inwards.'

There is perhaps no other concept which causes quite so much confusion for the student of mechanics as that of 'centrifugal force'. In order to avoid or remove such confusion it is essential that students understand the characteristics of force, as specified in chapter 1, section 1. In particular, applied forces which appear on the left hand side of $\mathbf{F} = m\mathbf{a}$ are caused by other bodies. Here there is no other body causing a force of magnitude '$mr\omega^2$'.

Conclusion

$mr\omega^2$ is not a force in the Newtonian sense.

The student's comment that a force is needed to stop the mass falling inwards arises from an erroneous idea that the mass is in equilibrium! If the string and mass were at rest in the plane then a force **F** would be needed to hold it outwards – and so prevent it falling inwards. In such a case the mass is in static equilibrium, having no acceleration!

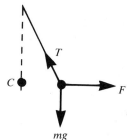

When the mass rotates about the vertical however, it accelerates towards the centre of the circle, C, under the action of a resultant force acting towards C. As we have seen, this force is the resultant of **T** and $m\mathbf{g}$.

We recommend the following approach to the solution of dynamics problems.

---

1 Force diagrams should include only applied Newtonian forces, in other words, forces caused by other bodies.
2 These Newtonian forces constitute the **F** in Newton's second law $\mathbf{F} = m\mathbf{a}$ where **a** is the acceleration.

This approach is well illustrated in questions 4–6.

Conical pendulum, question 4

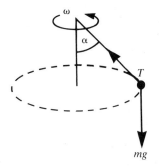

Fig. 4.5

Newtonian forces are tension **T** and weight *m***g**. Newton's second law in the vertical and radial directions yields

$$T\cos\alpha - mg = 0$$

$$T\sin\alpha = mr\omega^2$$

See 'Conical pendulum', chapter 9.

Fly on the turntable, question 5

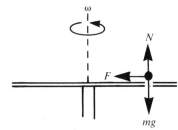

Fig. 4.6

The Newtonian forces acting on the fly are weight *m***g**, normal reaction **N**, and friction **F**, acting radially inwards.

Equations of motion in the vertical and radial directions are:

$$N - mg = 0$$

$$F = mr\omega^2$$

See 'Pennies on a turntable', chapter 9.

Marble in the groove, question 6

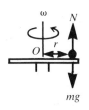

Fig. 4.7

Here there is no Newtonian force in the radial direction and therefore the marble cannot remain at rest relative to the turntable. (There is no force to produce an acceleration $r\omega^2$!) The marble must move relative to the turntable; r varies in time and we must include terms in the acceleration which take account of this variation.

In the radially outwards direction Newton's second law gives

$$0 = m(\ddot{r} - r\omega^2) \tag{4.1}$$

since in polar co-ordinates the acceleration of a particle radially outwards is $(\ddot{r} - r\dot{\theta}^2)$, and $\dot{\theta} = \omega$ in this case.

therefore

$$\ddot{r} - r\omega^2 = 0$$

which can be solved for $r(t)$ to give

$$r(t) = A\mathrm{e}^{\omega t} + B\mathrm{e}^{-\omega t}$$

where A and B are arbitrary constants. Clearly the radial distance of the marble from O increases exponentially with time.

Conclusion

These examples illustrate an approach to circular motion which is both straightforward and consistent. An observer, who is at rest, asks the question 'Which Newtonian forces are acting?' and these are included on a force diagram. Furthermore, there is no need to introduce the concept of centrifugal force!

Unfortunately this is not the end of the story; there are other difficult questions to answer relating to the experiences of an observer who is moving in a circle with the body. These are considered in References, Savage & Williams (7).

4.3 ACCELERATION

☺ **Correct answers**

7

(a) 'If speed is constant, then acceleration is zero!'
                    False!
(b) 'Average velocity cannot be zero over each rotation if its average speed is $a\omega$.'
                    False!

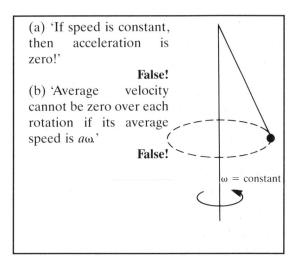

A car travels with steady speed along a winding road. Is its acceleration: zero ? No
constant ? No
variable ? Yes

If the car now travels along this winding road with increasing speed, insert the direction of its acceleration at A, B and C.

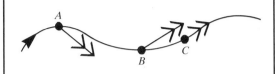

Students will generally come unstuck with questions 7, 8 or 9 if they fail to appreciate that speed, angular speed and angular acceleration are scalars whereas velocity, angular velocity and acceleration are all vectors. The most common misconception here is that:

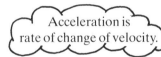

$$a = \frac{dv}{dt} \,!$$

So for example, in questions 7(a) and 8, some students will believe that if speed is constant then it follows that there is zero acceleration. This restricted view of acceleration is understandable since it is an intuitive idea of acceleration that students bring with them to a mechanics course. The teacher's task is to get the student to think carefully about the scientific definition of acceleration and what that means.

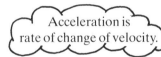

$$\mathbf{a} = \frac{d\mathbf{v}}{dt}$$

It will then become clear that a body's acceleration might involve either a change in speed or a change in the direction of motion or both. Consider for example two satellites A and B launched by the space shuttle high above the Earth.

Ignoring the effect of the sun, moon and other planets, the only force acting on each is the Earth's gravity acting radially inwards which produces an acceleration $\mathbf{a}$ in that direction.

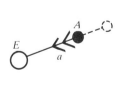
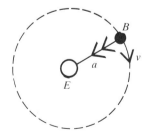

Fig. 4.8

Satellite A is launched from rest and therefore accelerates radially inwards with increasing speed:

$$a = \frac{dv}{dt} \qquad \begin{array}{l} \text{– speed varying} \\ \text{– direction constant.} \end{array}$$

Satellite B is launched so as to orbit the Earth in a circle of radius r, with a constant speed v, such that:

$$a = \frac{v^2}{r} \qquad \begin{array}{l} \text{– speed constant} \\ \text{– direction varying.} \end{array}$$

Different initial conditions cause these two satellites to follow different paths, yet both accelerate towards the Earth as indicated by Newton's second law.

Curiously, students often have more difficulty with these concepts in one-dimensional motion than they do in two dimensions.

(a) 'You have put $\ddot{x}$ downwards: surely this is incorrect since it is really accelerating upwards.'

(b) 'How can a length x be *negative*?'

(c) 'The least speed is zero, which occurs when $\ddot{x} = 0$ according to calculus . . .'

Comment!

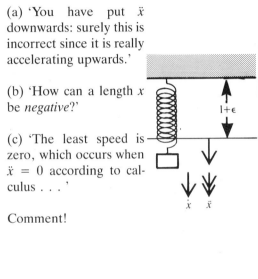

In question 9, it is *not* strictly accurate to say that x, $\dot{x}$ and $\ddot{x}$ are the displacement, velocity and acceleration of the mass! In fact, if $\mathbf{i}$ is the unit vector in the downward direction, then displacement $\mathbf{r} = x\mathbf{i}$
velocity $\mathbf{v} = \dot{x}\mathbf{i}$
and acceleration $\mathbf{a} = \ddot{x}\mathbf{i}$.

Now x, $\dot{x}$ and $\ddot{x}$ can be seen to be the **components** of the vectors $\mathbf{r}$, $\mathbf{v}$ and $\mathbf{a}$ in the vertical direction. Note that they may be positive or negative.

Correct answers

The following comments may be given in response to the stated misconceptions.

'You have put $\ddot{x}$ downwards: surely this is incorrect since it is really accelerating upwards.'

Comment: The vector $\ddot{x}\mathbf{i}$ *will* be upwards if $\ddot{x}$ turns out to be negative.

'How can a length x be *negative*?'

Comment: x is not a **length**, but a number. The length of the displacement vector $x\mathbf{i}$ is $|x|$.

'The least speed is zero, which occurs when $\ddot{x} = 0$ according to calculus . . .'

Comment: It is not true that $\ddot{x}$ is the derivative of **speed**, because speed is $|\dot{x}|$ which is not the same as $\dot{x}$.

4.4 NEWTON'S THIRD LAW

Being able to quote Newton's laws correctly is certainly no guarantee that one can correctly apply them in practice. This is particularly true of the third law as illustrated in question 10 where a block of mass m rests on a rough table.

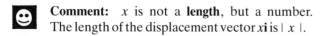

10

A light and inextensible string is attached to the block as shown. Weights are gradually added to the holder and the block of mass m remains at rest subject to the forces shown.

What is the relationship between F and N?

$$F = \mu N!$$

Insert the force(s) exerted by the block on the table.

Many students will correctly assert that 'the table exerts a force on the block and the block exerts an equal and opposite force on the table!' and then proceed to argue that since the weight of the block is $m\mathbf{g}$ then the force exerted on the table is $m\mathbf{g}$ as shown, rather than the correct answer.

Correct answers

$F \leq \mu N$

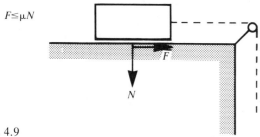

Fig. 4.9

Students may need reminding that the force on the table (due to the block) arises from an interaction with the block which produces equal and opposite forces of interaction. Only in the special case of a block freely resting on the table (with no string attached) is this force equal in magnitude to mg. (See also question 14, 'What do bathroom scales measure?')

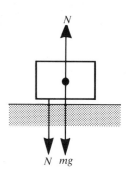

4.5 FRICTION

Of the various contact forces, friction is the most elusive. It is

■ a force which both opposes motion and yet makes motion possible!
■ a force whose action is easily misunderstood!

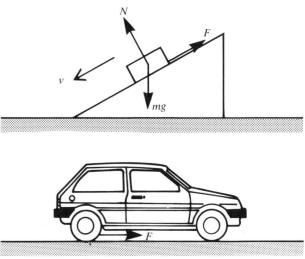

Fig 4.10

Question 10 asks for the relationship between friction F, and normal reaction N; namely

$$F \leq \mu N$$

There are two points to note here. Many students are apparently unaware of the inequality and simply quote the relation between limiting friction F_L and normal reaction

$$F_L = \mu N. \qquad (4.2)$$

This is what one might expect from students who have never actually done or seen the experiment in which weights are added to the weightholder until the block of mass m begins to slide. (See 'The law of friction', chapter 7.)

Secondly, many students are unaware of the real status of equation (4.2) – that it is simply an experimental or modelling law, only approximately valid over a limited region $0 \leq N \leq N_0$. (See chapter 2, section 3.)

Correct answer, question 11

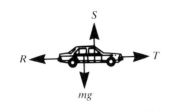

A car travels at constant speed along a straight and horizontal road. If the car experiences a force due to air resistance R, what other force must act on the car?

It is usually argued that the resultant force on the car is zero since it travels with constant velocity **v**. Therefore air resistance **R** must be balanced by a force **T**, of equal magnitude acting in the opposite direction. **T** is often referred to (by students) as the 'engine force' or 'traction'. Very few realise that traction is produced by friction exerted by the ground on the tyres. The engine simply causes the axle to rotate and if there were no friction there would be no forward motion!

 A common misconception is highlighted in the following remarks:

'How can traction be produced by friction?'
'Friction acts in the opposite direction to motion, doesn't it?'

It is a misconception to assume that if a car goes forward then friction must necessarily act in the opposite direction. One must look carefully at the point of contact between tyre and ground.

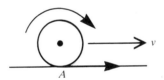

If the engine causes the axle and wheel to rotate in a clockwise sense, there is a tendency for the wheel to skid to the left at the point of contact, A. Friction comes into play to oppose the relative motion between the skidding tyre and the ground and acts to the right as shown above.

A car moves in a circle at constant speed subject to air resistance of magnitude R. Indicate by an arrow the direction of the friction force exerted by the road.

 Question 12 draws attention to another misconception:

'Friction always acts either in the direction of motion (as traction as in question 11) or in the opposite direction (when the body slides, as in question 10).'

The car travels in a circle and therefore the resultant force must be towards the centre of the circle. This force is the resultant of air resistance **R**, and friction **F**. Therefore **F** must be directed as shown.

 Correct answer, question 12

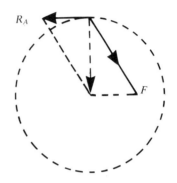

Fig. 4.11

A snooker ball rolls down an inclined plane and experiences weight **W**, normal reaction **N** and a frictional force. **F**. What can you say about its total energy: kinetic and potential?

 Question 13 identifies yet another misconception regarding friction:

'Friction is always dissipative – reducing the total mechanical energy of the body.'

Indeed it is true that friction is dissipative if the ball either slides or skids. However, if the ball rolls then friction is either absent or it is present but is non-dissipative.

Since the ball rolls, there is no relative motion between the ball and the ground at the point of contact A, that is, the ball is instantaneously at rest at A and therefore friction does no work and the total energy of the ball is conserved. (See 'Energy of a rolling ball' chapter 12.)

 Correct answer, question 13

Total energy of the ball is conserved.
Kinetic energy + potential energy = constant.

4.6 WEIGHT AND WEIGHTLESSNESS

 What forces do bathroom scales measure?

(See 'Bathroom scales and a brush', chapter 5.)

 'Bathroom scales measure the weight W of a body.'

 Correct answer

'Bathroom scales measure the reaction between the scales and the body.'

Note that:

1 the body exerts a force **N** on the scales and the scales exert an equal and opposite force on the body.
2 the weight, W, of a body of mass m (on the Earth) is the gravitational force exerted upon it by the Earth): $W = mg$.
3 N may be the same as W. It may also be larger or smaller than W as illustrated in the above-mentioned practical investigation or in question 15.

 Do the scales read heavier or lighter when you are in a lift accelerating upwards?

 Correct answer

'The reading on the scales is greater'.
'The scales will read heavier.'

This follows from Newton's second law applied in the direction of the acceleration a.

The magnitude of the reaction N is greater than mg and so the scales will read heavier!

$N - mg = ma$
$N = mg + ma$

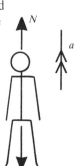

Fig. 4.12

A similar analysis will show that a suitcase you are holding will feel heavier when you are accelerating upwards in a lift,

$$P > mg!$$

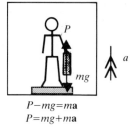

$P - mg = m\mathbf{a}$
$P = mg + m\mathbf{a}$

Fig. 4.13

 A woman is in free fall holding her suitcase. Why does it feel weightless when air resistance is negligible?

 Correct answer

The correct answer is obtained by applying Newton's second law in the direction of free fall.

$$W - P = mg$$

But

$$W = mg$$

therefore

$$P = 0.$$

Fig. 4.14

No force is required to support the suitcase, which therefore feels weightless!

Another common misconception is to assume that:

'weightlessness means the absence of weight'

In fact 'weightlessness' usually means an absence of reaction between the body and that with which it is in contact. For example, if a woman stands on bathroom scales and experiences weightlessness there is no reaction between her and the scales, $R = 0$. This would be the case if both were in free fall!

Fig. 4.15

Note, however, that the woman's weight W is not zero – Earth's gravity still acts and $W = mg$.

 An astronaut orbiting the Earth in a spaceship feels weightless. Why?

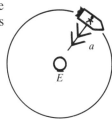

Once again, the correct answer is found by applying Newton's second law to both the astronaut (mass m) and the spaceship (mass $\overline{M}$) which are assumed to orbit the Earth in a circle of radius r, with an acceleration **a** directed towards the centre of the Earth.

$$m\,\frac{MG}{r^2} = ma$$

$$\overline{M}\,\frac{MG}{r^2} = \overline{M}a.$$

The applied force in each case is given by Newton's gravitational law, chapter 1, section 1 and it follows that

$$a = \frac{MG}{r^2}$$

Now if there were a reaction R between the astronaut and the spaceship then the above equations would become

$$m\,\frac{MG}{r^2} - R = ma$$

$$\overline{M}\,\frac{MG}{r^2} + R = \overline{M}a$$

which give different values for a.

Hence R is identically zero, and no reaction implies 'weightlessness' for the astronaut in the orbiting spaceship!

 Correct answer

'The astronaut feels weightless because there is no reaction between herself and the spaceship!'

PART B

Practical investigations

5

INVESTIGATING FORCE

This chapter contains a number of practicals which motivate the use of force diagrams when modelling bodies in equilibrium or which show how forces can be combined and resolved.

Bathroom scales and a broom shows how Newton's first and third laws can be applied, using labelled force diagrams to explain the reading on the bathroom scales in various situations.

Equilibrium is intended to provoke discussion of the concept of equilibrium, and the assumptions which are made when modelling with smooth pulleys and light strings.

Lines of force provides a practical which illustrates the key ideas involved when a body is in equilibrium subject to two, three or four forces.

Spring balances (1) is both a simple practical and a modelling investigation concerned with force in one dimension.

Spring balances (2) / parallelogram law provide practicals for validating both the parallelogram law of forces and the resolution of a force into components.

Three masses can be used with GCSE or A-level pure mathematics students to provide data for modelling with functions and graphs.

Investigating force with bathroom scales and a broom

Aims

- To discuss force, particularly **weight** and **reaction**.
- To appreciate that scales measure reaction, which is not always the same as weight.
- To introduce and practise the use of force diagrams to solve problems or explain events.
- To discuss Newton's laws in simple contexts.

Equipment

Bathroom scales or Newton's scales, broom (or pole), plank.

Plan

An example of a lesson plan for 1 hour 10 minutes

(Possibly a 'taster' course for post GCSE students or an introduction to force for A-level students.)

Introduction (10 minutes)

With the aid of kilogram scales and a Newton scale.

- 'Guess my weight', '. . .a volunteer's weight', '. . .how big a reading can you get by squeezing with your hands?'
- What **force** do the scales measure? What are the appropriate units? How do kilograms and Newtons compare? Can I reduce my 'weight'?
- What will happen if I press down on the floor with a pole? How (or by how much) will the reading change?
- Make a list of interesting 'What if I. . .' on the blackboard.

Group work (20–30 minutes)

Direct the class (in suitable groups) to write down for each case they try,

(i) their guess,
(ii) the result,
(iii) their explanation.

(Their guess will usually be correct: it's the explanation which is crucial.) Provide the first worksheet for stimulation of ideas.

At this stage, try not to give too many answers or to repeatedly 'correct' their language. This may inhibit them and stem the flow of ideas.

Each group needs at least one set of scales and a broom. This is your chance to find out what they know about force and Newton's laws. Get those who have ideas to explain them to those who do not.

Class discussion and force diagrams (20–30 minutes)

Take a report of each group's 'solution' to the problem of 'reducing the weight by pressing down with the

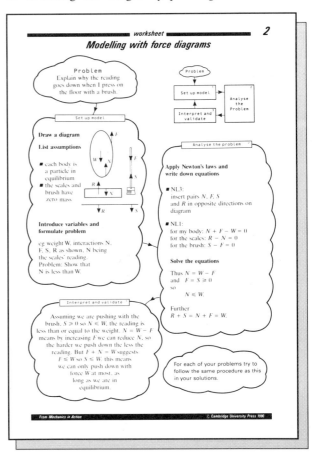

broom', without comment. Differences in the reports should provide fuel for discussion and motivate your exposition. Work through the force diagram solution (on worksheet 2), stressing the key ideas and contrasting them with misconceptions that may have arisen.

·**Key ideas are:**

- the distinction between 'weight' and 'scales reading'.
- separated force diagrams.
- Newton's first law.
- Newton's third law.

Choose another example or two for close study if there is time.

Follow up

Individuals should now provide some solutions, using the 'Modelling with force diagrams' approach, to the situations they looked at earlier. The second worksheet provides some guidance for this.

Solutions

Solutions are presented following the 3-stage modelling process.·

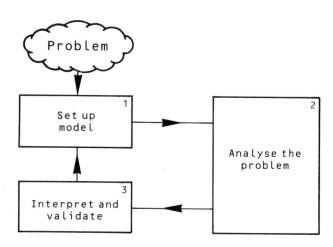

In these problems, stage 2 is short and simple. Most of the work is in stage 1 – in the drawing and labelling of good force diagrams. Occasionally it will be of interest to go back and **refine the model**. For instance, in the worked example, we may wish to take account of the mass of the broom and/or the mass of the scales, to make the model more realistic!

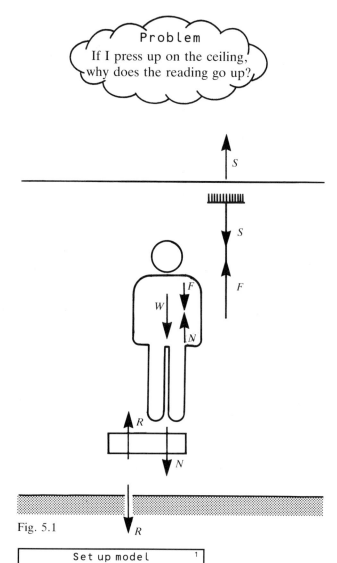

Fig. 5.1

| Set up model | 1 |

(i)   See Fig. 5.1.
(ii) Assumptions as in the worksheet.
(iii) Introduce W, R, N, F, S as shown.

| Analyse the problem | 2 |

Problem: to show that N is greater than W!
Use Newton's Laws (NL)

NL3: R, N, F, S introduced in pairs.
NL1:   for my body $N-F-W = 0$
           for broom $\quad F-S \quad = 0$
           solution $\qquad N = W+S$

therefore

$$N \geq W.$$

| Interpret and validate | 3 |

Since $S > 0$ when I push up then $N > W$. The harder I press up, the larger is S and the greater is the reading N! Also the reaction experienced by the ground, R, is greater.

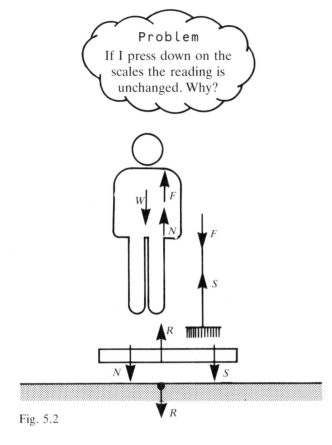

Fig. 5.2

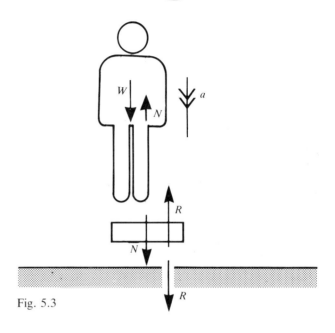

Fig. 5.3

| Set up model | 1 |

(i)   See Fig. 5.2.
(ii) As before.
(iii) Introduce W, N, F, R, S as shown. The scale reading is now $(N+S)$ or R but not N!

| Analyse the problem | 2 |

Problem: to show that $N+S=R=W$ no matter how big F or S is!

NL3: N, F, R, S are introduced in pairs.
NL1:   for my body $N+F-W = 0$
       for the broom $S-F = 0$
       for the scales $R-N-S = 0$

       solution $N+S = R = W$.

| Interpret and validate | 3 |

Although N may reduce to zero if F is large enough (i.e. if $S = W$) the scale reads $(N+S)$ which remains unchanged.

| Set up model | 1 |

(i)   See Fig. 5.3.
(ii) Assuming my body is now a particle **accelerating** downwards with acceleration **a**.
(iii) Introduce W, N, R.

| Analyse the problem | 2 |

Problem: find N in terms of a!

NL3: N, R, introduced in pairs.
NL1:   for my body $W-N = ma$
       weight $W = mg$

solution $N = W-ma = m(g-a)$.

| Interpret and validate | 3 |

If $a \geq 0$; $N < mg = W$, so the reading is **less** than the weight.

   Therefore while accelerating downwards the reading **falls**. However, when $a \leq 0$, accelerating upwards the reading is **greater** than my weight. This is the lift problem – question 15 on the questionnaire in chapter 4.

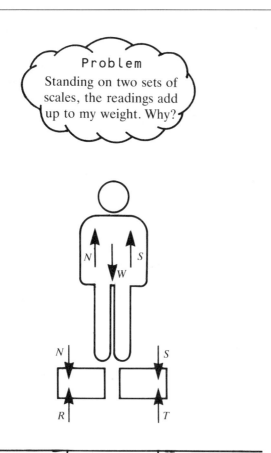

Problem
Standing on two sets of scales, the readings add up to my weight. Why?

Fig. 5.4

This is identical to the worked example on the worksheet 'Modelling with force diagrams', except that here we have a leg instead of a broom, and it presses on scales instead of the floor.

$$N+S-W = 0$$
so $N+S = W$ or $N = W-S$ or $S = W-N$.

The readings add up to my weight; I can increase one reading only at the expense of the other. Neither reading can be **more** than my weight because neither can be **less** than zero!

Key ideas and misconceptions

Students begin their sixth form mechanics course with many misconceptions about force, as the following examples show. If you ask students at this stage to give examples of forces, some will include mass, acceleration, air pressure, and even energy.

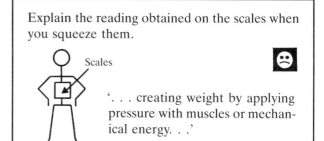

Explain the reading obtained on the scales when you squeeze them.

'. . . creating weight by applying pressure with muscles or mechanical energy. . .'

The purpose of this practical is to encourage students to discuss their ideas of force in the context of a simple, familiar situation, so that you can correct any major misconceptions revealed in the process. We outline here the key areas of interest and concern. But do not be surprised if your students produce others. . . Intuitive mechanics throws up many curious and beautiful formulations, treasure them!

1 'Weight' and the reading on the scales

In common everyday English usage the 'weight' of a body is used to refer to the force of the body on the floor on which it rests. Even scientists and engineers may use the term in such a way, for example when they describe an orbiting astronaut as 'weightless'. These misconceptions therefore arise frequently in students' work.

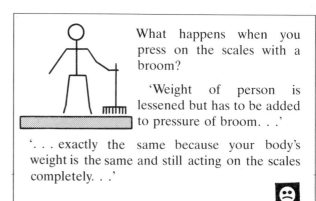

What happens when you press on the scales with a broom?

'Weight of person is lessened but has to be added to pressure of broom. . .'

'. . . exactly the same because your body's weight is the same and still acting on the scales completely. . .'

It is important to introduce the correct technical definition of a body's weight, W, as:

'the force of gravitational attraction applied by the Earth to the body'.

This is to be clearly distinguished from the forces acting on the scales which cause the indicator to deflect. The scales actually read a value which is designed to be proportional to the total downward force applied to its upper surface.

As the various solutions show, the reading on the scales varies, but W does not! (See chapter 4.)

2 Force diagrams and the concept of 'applied force'

Students commonly draw diagrams and insert forces in a confused way, lacking clarity about which force applies to which body.

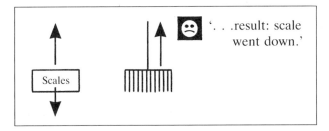

'. . .result: scale went down.'

It will help if all bodies involved are clearly separated in a separated-body diagram so that it is clear which body a particular force acts on.

An essential point about all Newtonian forces is that they are external, applied forces. A Newtonian force acting **on** a body, A say, is due to or caused **by** a distinct body, B. We call this the force of B on A:

■ the force of me (B) on the scales (A) is the vector **N**.

■ the force of the Earth (B) on me (A) is the vector **W**.

■ the force of the brush (B) on me (A) is the vector **F**.

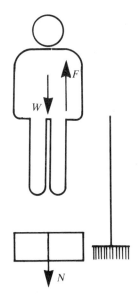

Fig 5.5

Every force in Newtonian mechanics can be attributed to the action of another body. If there is no such body, the 'force' is probably an erroneous invention! (See centrifugal force and engine force in chapter 4).

3 Newton's third law (NL3)

Students usually have problems with Newton's third law because they do not appreciate that the two forces referred to always act on different bodies (see Newton's third law, chapters 1 and 4). If you ask your students to name two 'equal and opposite' forces due to NL3,

they will often name two forces acting on the same body which are equal and opposite due to Newton's first law (NL1). For example, when a person stands on the scales, weight is equal and opposite to the reaction of the scales. This is because the body is in equilibrium and **not** because of NL3, which applies whether or not the bodies involved are in equilibrium.

In fact, NL3 states that

'whenever two bodies A and B interact, they exert equal and opposite forces on each other. So if A acts on B with force **F**, B acts on A with force $-$**F**'.

This, of course, applies to **all** Newtonian forces, which should therefore normally be drawn on a force diagram in pairs:

■ with the force of me on the scales, **N**, we also insert the force of the scales on me, $-$**N**.

■ with the force of the Earth on me, **W**, we can also insert a force of me on the Earth, $-$**W**, but we are not concerned with the force diagram for the Earth, so we don't usually include $-$**W**.

■ with the force of the broom on me, **F**, we also insert the force of me on the brush, $-$**F**.

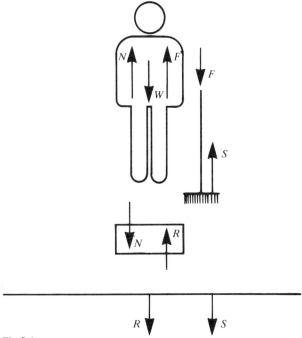

Fig 5.6

Similarly, we can insert the pairs of interactions **S** and $-$**S**, and **R** and $-$**R**.

4 Equilibrium and Newton's first law (NL1)

Most of the examples investigated involve the concept of equilibrium, which is defined in Newton's first law (NL1).

'A body which has no resultant force acting upon it is in equilibrium and so, if stationary, it will remain stationary, but if moving with velocity **V** it will continue to move with the same velocity **V**.'

Therefore, if a body is in equilibrium under the action of two forces, their resultant must be zero and they must be equal in magnitude and opposite in direction. Under three forces, we have a more interesting result $\mathbf{F}_1+\mathbf{F}_2+\mathbf{F}_3 = \mathbf{0}$: see 'Lines of force' later in this chapter for more practical work on this topic.

However, it is the concept of equilibrium which is most difficult. Many students see equilibrium as static equilibrium, acceleration as movement, and force as a cause of motion (see chapter 4). In fact, force causes **change** of motion, which is change of $\mathbf{V}$. It is difficult to remedy this misconception without considering dynamic equilibrium, where $\mathbf{V}$ is constant but not zero. So in this case you need to take your scales for a ride in a lift. If no lift is available, a thought experiment will provoke discussion: what happens to the reading if you go up or down in a lift?

'Going up in the lift, it increases. You are pushing against gravity. Going down, it decreases, moving with gravity.'

'When going down you are going with gravity. . .
So less gravity while moving.'

The reading on the scales is the same when moving **uniformly** with constant speed in the lift as it is when at rest. Indeed, one is quite unaware of the motion except during the acceleration of the lift at the beginning and end of the ride.

Extensions

- walking the plank.
- leaning against a wall.

Equilibrium

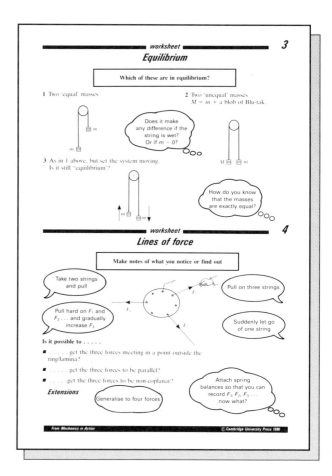

Aims

- To discuss the concept of 'equilibrium'.
- To introduce modelling, including the notions of smooth pulleys, light strings and connected particles (see chapter 2 on modelling).

Equipment

A 'smooth' pulley, and masses are provided by Unilab in the Leeds Mechanics Kit or in the Practical Mathematics Kit.

Plan

Look at each situation on the worksheet with the class as a whole. **You** hold the pulley and the string, and say 'What will happen when I let go?' Get everyone to guess.

'Can you justify your guess? What is your argument?' Try to exploit the different opinions of each different student. If necessary take a contrary position yourself. 'Are you certain? Can you prove it? What are your assumptions?' are good questions to provoke thought.

Do not let them see what happens until you have squeezed every drop of juice from the argument: it's too easy to 'explain' after the event!

Solution to 1

The solution depends on the relative effect of

(a) the weight of the string,
(b) the friction in the pulley bearing,
(c) the accuracy of the two 'equal' masses.

Assumptions 1

Smooth pulley bearing, heavy string. Consequently, the forces on the pulley are unequal and the pulley rotates, accelerating the left hand 'particle' downwards.

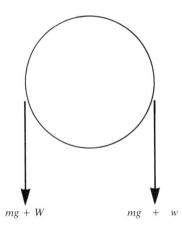

$$mg + W \qquad mg \quad + \quad w$$

W and w are the weights of the two different lengths of string

Assumptions 2

Smooth pulley bearing, light string, equal masses: the result is equilibrium.

Assumptions 3

Frictional pulley bearing, heavy string and/or unequal masses. The result is determined by whether or not the difference in masses is sufficient to overcome the friction due to the pulley and cause acceleration.

☹ Misconceptions

Two points about equilibrium may arise.

1 Unaccountably (perhaps by analogy with a see-saw motion) some students expect the right hand mass to go **down** when released!
2 Equilibrium is often thought to be incompatible with motion, that is 'equilibrium' is equated with 'static equilibrium'. Use every opportunity to disrupt this idea.

Extensions

In addition to the situations suggested in the worksheet, the modelling assumptions can now be explored further by the following:

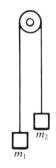

unequal masses over a 'rough' peg

(a)

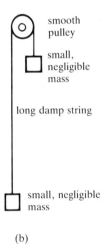

smooth pulley

small, negligible mass

long damp string

small, negligible mass

(b)

smooth pulleys

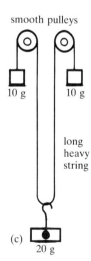

10 g   10 g

long heavy string

(c)   20 g

Fig. 5.7

Lines of force

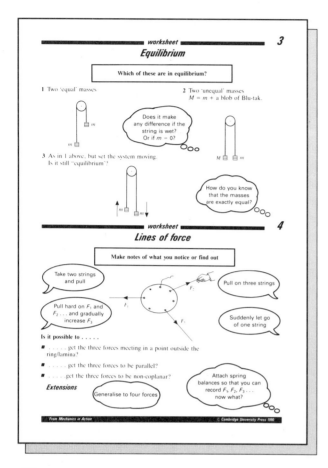

Aims

- To discuss the action of two, three or four forces on a body in equilibrium.
- To obtain the principal results about lines of action.

Equipment

A ring or wooden lamina with holes to fix strings. String, paper and pens. 3 or 4 arms.

Fig. 5.8

Plan

In groups or as a class, discuss the questions raised in the worksheet. This can be as little as 15 minutes' work.

Lines of action marked on a large sheet of A2 paper can be drawn and shown to act (roughly) through a point. This can be done also at a blackboard with chalk, whose thickness is usually enough to blur errors!

Errors will be particularly prominent when the lines are nearly parallel, as in Fig. 5.8. Four forces do **not** usually meet in a point, and the case of 3 parallel forces needs pointing out.

A fourth 'hidden' force can be introduced sneakily by relaxing the 3 strings as shown in Fig. 5.9. The fourth force is of course the weight of the lamina!

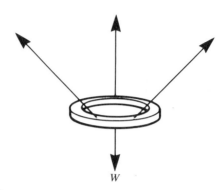

Fig. 5.9

Solutions

> **Problem**
>
> To investigate three forces in equilibrium.

| Set up model | 1 |

Assume the lamina has negligible weight and there is no friction.

Model the three forces as vectors, $\mathbf{F}_1$, $\mathbf{F}_2$ and $\mathbf{F}_3$.

| Analyse the problem | 2 |

Apply Newton's first law

$$\mathbf{F}_1 + \mathbf{F}_2 + \mathbf{F}_3 = \mathbf{0}$$

or

$$\mathbf{F}_3 = -(\mathbf{F}_1 + \mathbf{F}_2)$$

so, for example $\mathbf{F}_3$ is the inverse of the vector sum of $\mathbf{F}_1$ and $\mathbf{F}_2$. Also, the line of action of $\mathbf{F}_3$ is the same as the line of action of the resultant of $\mathbf{F}_1$ and $\mathbf{F}_2$.

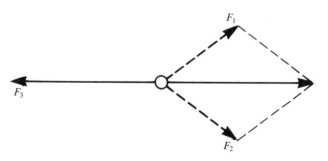

Fig. 5.10

Interpret and validate [3]

When F_3 is gradually increased from zero the angle between F_1 and F_2 is gradually decreased from 180°.

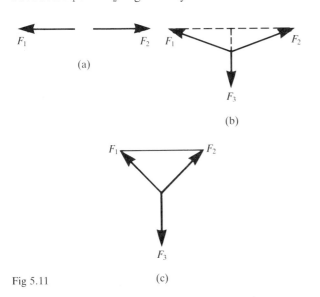

(a)

(b)

(c)

Fig 5.11

When two strings are pulled, they line up in a straight line.

When three strings are pulled the lines are coincident, although some error will be observable because of friction and the weight of the lamina. These effects can be minimised by increasing the tension in the strings and by flicking or lifting the lamina a few times before marking the points.

Comments

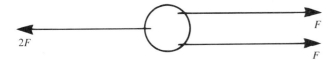

- The three forces can be parallel.
- The forces can be non-coplanar if the weight of the lamina is significant.
- Spring balances must be zeroed. Please check you have good spring balances and calculate their accuracy and reliability before using them. You can do this by attaching a few masses and checking their readings. Also, you can check quickly that they agree:

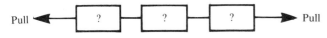

Extensions

Generalise to four forces.
Attach spring balances so that you can record F_1, F_2, F_3 . . . now what?

Spring balances (1)

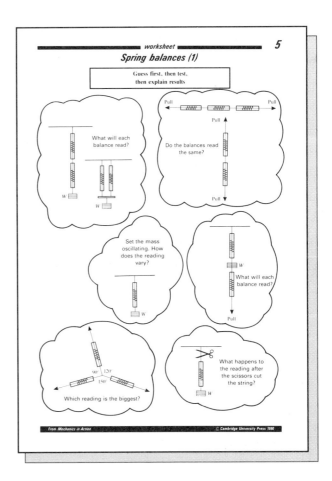

Aims

- To provoke discussion of force in one dimension.
- To analyse situations involving tension, providing a check using spring balances.

Equipment

Good spring balances, masses and some string. A pencil can be used to attach two spring balances in parallel.

Plan

Provide each group with three balances, string, a mass holder and set of 100 g masses. Spring balances reading 0–10 N and four or five 100 g masses work best. Check that the spring balances are zeroed and reliable before the class disprove Newton's laws for you!

Solutions

Set up model [1]

Assume bodies are in equilibrium in each case.
Assume spring balances are accurate.
Assume there is no friction.

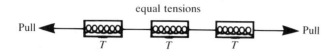

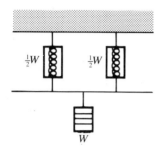

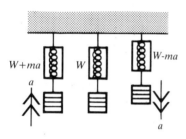

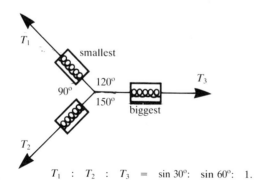

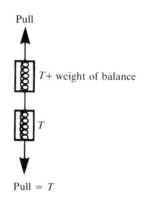

Pull

$T+$ weight of balance

T

Pull $= T$

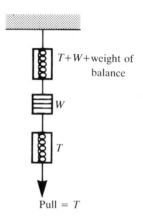

$T+W+$weight of balance

W

T

Pull $= T$

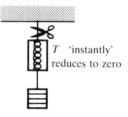

T 'instantly' reduces to zero

$T_1 : T_2 : T_3 = \sin 30^\circ: \sin 60^\circ: 1.$

Fig. 5.12

🙁 *Misconceptions*

What happens when the string is cut relates to 'free fall' and question 16 on the questionnaire in chapter 4.

Extension

The last situation leads naturally into 'Spring balances (2)'.

Spring balances (2)

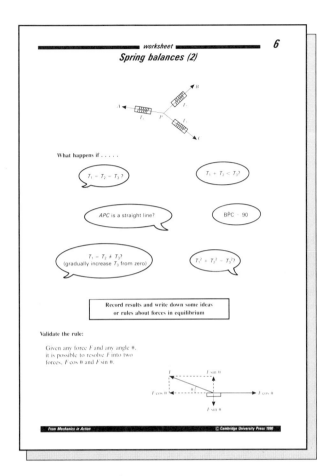

Solutions

$T_1 = T_2 = T_3$ gives
$A\hat{P}C = C\hat{P}B = B\hat{P}A = 120°$.

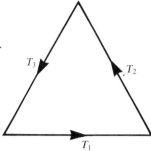

It is impossible to get $T_3 > T_1 + T_2$, just as it is impossible to get a third side of a triangle longer than the sum of the other two sides.

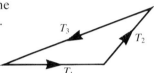

If APC is a straight line, then B lies on APC and $T_1 = T_2 + T_3$ or $T_3 = T_1 + T_2$ or $T_2 = 0$. If $T_1 = T_2 \neq T_3$ then we have two angles the same.

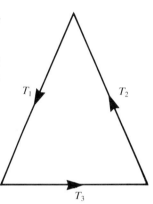

Aims

- To introduce (or verify) the parallelogram or triangle of forces.
- To verify the rule for taking components of a force, that is, resolving in two directions.

Equipment

Three good spring balances and large sheets of poster paper for each group.

Plan

Spring balances are only of limited accuracy: check your equipment. If the equipment is not very good you will end up teaching your students a sound experimental principle: your results are only as accurate as your measuring instruments. You can always switch to 'Three masses' if you are dissatisfied with errors.

If $T_1^2 + T_2^2 = T_3^2$ then we have a right angle :

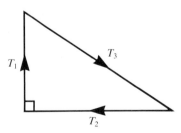

Three masses

worksheet ━━━━━ 7
Three masses

Problem
What is the connection between masses A, B, C and angle D?

Start with $A = C$ and vary B.

Describe what happens as you increase B or decrease B.

Why have $A = C$? Why not try $A = B = C$? Or why not fix angle D, say 90 degrees, and try to get values of A, B and C?

Try any values of A, B, C. Guess what will happen if . . .
– you increase A
– you increase B
– you increase C
Were you right?

Suppose
$B > A + C$?
$C > A + B$?
$A > B + C$?

From Mechanics in Action   © Cambridge University Press 1990

Aims

- To investigate the relationships between four variables.
- To explore (or apply) the theory of three forces in equilibrium.

Equipment

Good 'smooth' pulleys and a range of masses are provided by Unilab in the Leeds Mechanics Kit or in the Practical Mathematics Kit. A range of 100 g, 50 g and 10 g masses are required.

Plan A

GCSE or A-level pure mathematics students can make use of this investigation to use functions and graphs to model 'simple' empirical data compiled in tables. Variables have to be systematically controlled and tabulated. Data is entered on a graph, preferably a function graph plotter or graphic calculator, and functions found to fit the data.

For example if $A = C$, then we get $B = 2A \cos(\frac{1}{2}\hat{D})$. The graph of B against $\hat{D}$ looks like Fig. 5.13. A graph plotter such as FGP will plot $B = 2A \cos(\frac{1}{2}\hat{D})$ and allow variation of the constant A until the graph fits the plotted points.

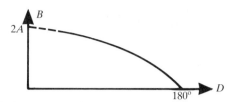

Fig. 5.13

Plan B

Set up the three masses as shown on the worksheet and discuss the questions as a group.

'Guess what happens if I take a 10 g mass off B' etc.

This should lead to an improved intuition of the three forces in equilibrium. Precise measurement, drawing and verification of the parallelogram law follows.

Solution

Set up model	1

Assume there is no friction in the pulleys.
Assume the string is light and the masses are accurate.
Assume 'the system' is in equilibrium.

Analyse the problem	2

If $A = C$, then $B = 2A \cos(\frac{1}{2}\hat{D})$.
If $A = B = C$, $\hat{D} = 120°$.
If $\hat{D} = 90°$, $A^2 + C^2 = B^2$.
$B > A + C$. This is impossible; equilibrium is broken, B accelerates downwards.
Similarly for $C > A + B$; C accelerates downwards.
In any particular configuration increasing one mass (say A) leads smoothly to a new equilibrium position as long as the mass A lies between $(B-C)$ and $(B+C)$.

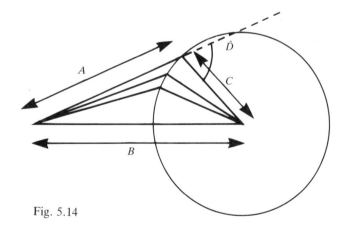

Fig. 5.14

Interpret and validate	3

Good results are usually obtained provided the effects of friction are small compared with the three weights.

Extensions

See 'Parallelogram law'.

Parallelogram law

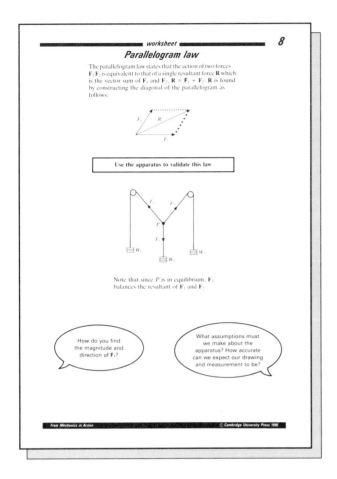

■ worksheet ■ 8

Parallelogram law

The parallelogram law states that the action of two forces F_1 F_2 is equivalent to that of a single resultant force R which is the vector sum of F_1 and F_2. $R = F_1 + F_2$. R is found by constructing the diagonal of the parallelogram as follows:

Use the apparatus to validate this law

Note that since P is in equilibrium, F_3 balances the resultant of F_1 and F_2.

How do you find the magnitude and direction of F_3?

What assumptions must we make about the apparatus? How accurate can we expect our drawing and measurement to be?

From Mechanics in Action   © Cambridge University Press 1990

Pairs of students might take data, each for a different set of three forces, and check by drawing that the parallelogram law works.

Experience shows that results are good to within one or two degrees. A little care must be taken to avoid parallax errors when marking positions on the paper.

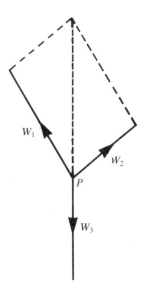

Important ideas are that the tension in each string equals the weight of the mass attached, and has nothing to do with the length of string! We assume that the pulley bearing is smooth, so there is no difference in tension across the pulley, and the strings are light, so there is no variation in tension along the strings (see chapter 2, sections 2.3a,b.)

Extensions

- Move P to one side and let go. Why does it always swing back towards the same equilibrium point P?
- Illustrate the resolution of a force by resolving tension T_3 into T_1 and T_2 at right angles as shown.

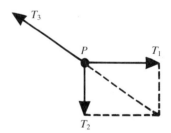

Aims

- To verify the parallelogram law of forces.
- To provoke discussion of modelling assumptions about the string, pulleys, masses.

Equipment

As for 'Three masses'.

Plan

Your strategy will be determined by the quantity of apparatus available. You need at least two good pulleys fixed to a board supported vertically on which you can mark points and draw the lines of force.

6
INVESTIGATING MOMENTS

This chapter contains a number of practicals and problems (modelling investigations) which involve the theory of moments and can be used in various ways in an A-level mathematics course.

Balancing a ruler provides a rich source of data which can be modelled with functions and graphs by pure mathematics A-level students. Alternatively it can be used as a modelling investigation which applies the theory of moments to obtain results which can be validated in practice. It also leads to the extended modelling investigation 'Design a roadblock'.

The law of moments / Beam balance is a practical investigation designed to validate the law of moments in simple situations.

Overhang consists of a simple practical and a challenging modelling investigation. The solution can be used to motivate a study of the series $\sum_{n=1}^{N} \dfrac{1}{n}$.

Centres of mass is a simple practical which can be used to motivate the theory for the centre of mass of uniform laminas or to apply the theory which has previously been taught.

The cable reel is a modelling investigation. Either a class demonstration or a practical investigation will reveal one or two surprises which are resolved by the law of moments.

The toppling tube provides an interesting and challenging mechanics problem which requires an appropriate model for the forces acting. It can be regarded as a starting point for further investigations.

Balancing a ruler

Aims

- To apply the theory of moments to a practical problem.
- To model data with a general law.

Equipment

Each pair of students should have at least one metre ruler, five 100 g masses and a 50 g mass and some Blu-tak. Ideally rulers of different mass should be available.

Plan

This practical has been tackled by different groups of pupils in various ways. GCSE pupils have collected data in a table, drawn a graph and used it to predict the balance point for different masses.

Lower sixth pupils have used it as an example of a moments calculation, and have validated Newtonian theory by making predictions and testing them.

Data collected has been modelled on a computer or graphic calculator by A-level students without the use of Newtonian theory.

The practical has also been used by upper and lower sixth A-level students to help them to understand an associated problem 'Design a roadblock using a given pole, road width and a suitable counterweighing system'.

The lesson plan and worksheet given is for a class who have met the theory of moments. The worksheet is intended to be sufficiently open for use in a variety of ways at different levels.

Example lesson plan for 1 hour 10 minutes with a lower sixth A-level class who have met the theory of moments

Introduction (15–20 minutes)

1 'A ruler normally balances at its midpoint. Why is this? Is it always true? What are we assuming?'
   - Assumptions are important and should be listed in any problem-solution.
   - The ruler is uniform!
2 Attach a 100 g mass with the aid of some Blu-tak. 'Where should it balance now? Roughly? Make a guess. . .'
   'Suppose it balanced at $d = 25$ cm, what would this mean?'
   'Suppose d is more/less than 25?'
   - If necessary, do the force analysis for an **example**, e.g. $d = 25$, then we could conclude that the mass of the ruler $M = 100$ g $= 0.1$ kg.
   - Now actually take a value for d by balancing the ruler on your finger.

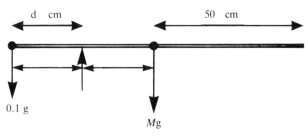

Fig 6.1. d cm is the distance to the balance point from one end of the ruler. M kilograms is the mass of the ruler and Mg Newtons is the weight of the ruler

'd is about 20 cm . . . how accurate is that? Can I get it to within 1 mm?'
'What can we say about the weight of the ruler? Calculate it now!'
- It is instructive to get a stated accuracy. Perhaps some will find a way of getting an accuracy to within 1 mm! (For example, resting on a wedge, or on the edge of the table.) It is important to attend to the accuracy of the answers, too. '100 grams ± 5 grams' rather than '102.3 grams' is desirable.

3 'Use mathematics to predict the balance point when 200 g is added: then check your theory in practice.'
'Make notes of all calculations and measurements.'

Group work, guided by the worksheet (35–40 minutes)

Groups of two can work well.

You may need to encourage lower sixth students to try several numerical cases before generalising.

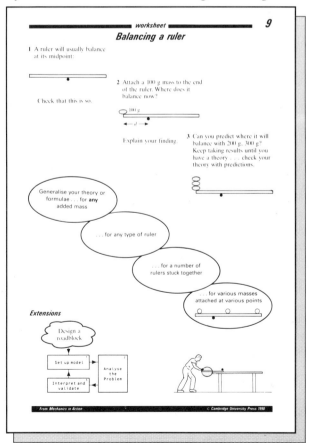

For example for $m = 200$ and $M = 150$,

$$d = \frac{150 \text{ g} \times 50}{200 \text{ g} + 150 \text{ g}}$$

and then by stages to something like

$$d = \frac{50\,Mg}{mg + Mg}, \text{ where } m \text{ is the mass attached}$$

or even

$$d = \frac{50 \times 150 \text{ g}}{n \times 100 \text{ g} + 150 \text{ g}},$$

where n is the number of 100 g masses.

In generalising, it is helpful to be able to contrast two or more such formulae from groups with rulers of different mass. You can even include a non-uniform ruler, a 30 cm ruler or a blackboard ruler to encourage the introduction of a further parameter, D say, the distance of the centre of mass from the end of the ruler.

Results are usually good. It is possible to get answers to within a few millimetres. Answers which are well out can be accounted for usually by not sticking the masses at the very ends of the ruler. Inaccuracy in measuring lengths can be significant. It is unlikely that the mass of the Blu-tak will really be significant because this is small compared to errors in the masses, which are usually only guaranteed to about 3%!

Class discussion (10–15 minutes)

Apart from intermittent discussion of the above points with groups or with the whole class, you may want to take them through the stages of modelling and ask them to present a report following the 3-stage modelling diagram.

Solutions

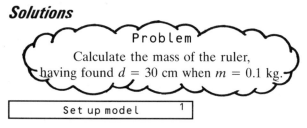

Problem
Calculate the mass of the ruler, having found $d = 30$ cm when $m = 0.1$ kg.

| Set up model | 1 |

The mass is a particle, the ruler is uniform, and exactly 100 cm long; my finger is a point. The ruler measures to within 2%, the mass is accurate to 2%. The mass of the Blu-tak is negligible.

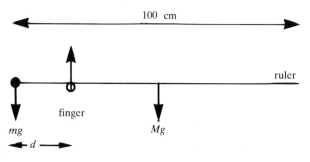

Fig. 6.2

| Analyse the problem | 2 |

For $m = 0.1$ and $d = 30$, the law of moments gives
$$0.1 \text{ g} \times 30 = Mg \times 20$$
$$\Rightarrow \qquad M = 0.15$$

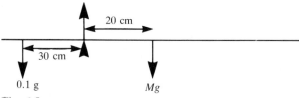

Fig. 6.3

| Interpret and validate | 3 |

The mass of the ruler is 150 grams, to within say 4% at worst; this is 150 ± 6 grams. Check this using kitchen scales.

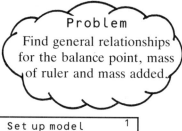

Problem
Find general relationships for the balance point, mass of ruler and mass added.

| Set up model | 1 |

As before.

| Analyse the problem | 2 |

The law of moments gives $mgd = Mg(50-d)$.
  Therefore

(i) $\qquad d = \dfrac{50M}{m + M}$,

or (ii) $\qquad M = \dfrac{md}{50 - d}$,

or (iii) $\qquad m = M\left(\dfrac{50}{d} - 1\right)$.

| Interpret and validate | 3 |

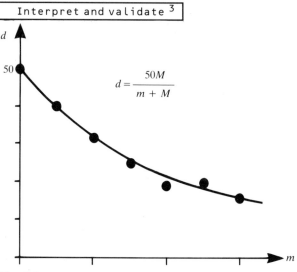

Fig. 6.4

(a) $d = \left(\dfrac{50M}{m+M} \right)$ (as shown in Fig. 6.4)

The balance point gets nearer the end as the added mass increases; eventually the mass of the ruler becomes negligible and the point of support approaches the end.

(b) The mass of the ruler can be found by adding a mass m, measuring d and calculating $M = \dfrac{md}{50-d}$.

(c) This is just the inverse function of that in (a).

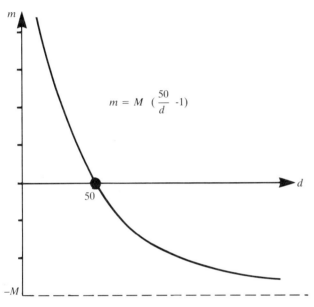

$$m = M \left(\frac{50}{d} - 1 \right)$$

Fig. 6.5

Any of these formulae or graphs can be used to make predictions and these should be tested in practice, providing further validation of the model.

There is special interest in the graphs of d against m for multiples of M, i.e. for rulers stuck together.

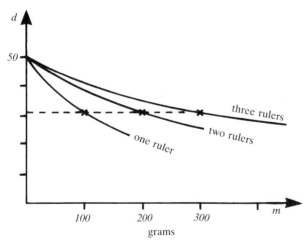

Fig. 6.6

For example, we find that they balance at the same value of d, when:

one ruler has 100 g attached,
two rulers have 200 g attached,
three rulers have 300 g attached.

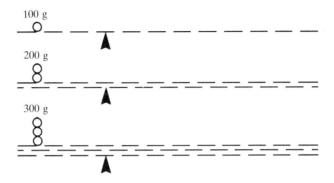

Extensions

1 A purely 'function and graphical modelling' approach can be taken, for example with a GCSE or pure mathematics/statistics A-level class.

Data can be collected in a table:

m	0	100	200	300	400
d	50				

and entered into a graph plotter (for example FGP on the BBC micro). This gives an opportunity to fit a function. Probably help will be required to get a sensible form:

- the function tends to x axis as m tends to ∞. What functions do you know that can do this?
- how can it start at (0,50)?
- would it be simpler to use a variable n where $n = m/100$?
- try something like $\dfrac{a}{n+b} \dots$!

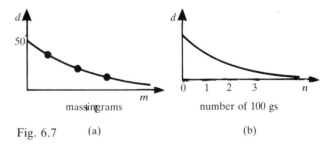

Fig. 6.7   (a)   (b)

2 An upper sixth A-level group might be required to generalise much more rapidly, going quickly to the extensions on the worksheet, and then tackling the open ended modelling investigation 'Design a roadblock'.

The law of moments / Beam balance

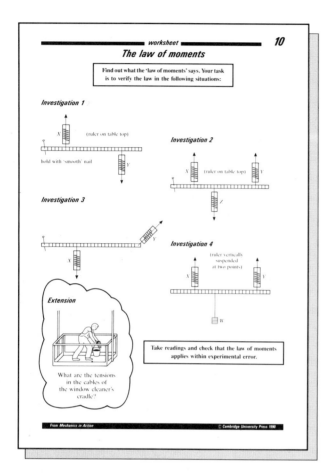

Aims

■ To introduce the law of moments in simple lever situations.
■ To verify the law in practice.

Equipment

A ruler with holes drilled or a strip of Meccano or Polymek. Good spring balances; 3 per group. A set of 100 g masses and mass holder.

Plan

Group work (30–40 minutes)

It is intended that students do these investigations with no help other than perhaps a statement of the law of moments. You may also wish to point out that investigations 1, 2 and 3 are to take place in a horizontal plane.

The idea is for them to investigate the consequences of the law of moments in each situation. It is not intended that they 'discover' the law; the important thing is that they are obliged to think the theory through.

Given a tight syllabus and timetable you might
(a) ask students to do this investigation in their own time and report next lesson.
(b) ask different groups to investigate different situations and report.

Solutions

Set up model	1

Assume
(i) there is no friction with the table top or at the hinges.
(ii) the balances accurately measure the forces applied to the ruler.
Introduce variables: X, Y, Z Newtons for the forces on the ruler. In investigation 4 let the attached weight be W Newtons and the weight of the ruler be Mg Newtons. Other variables x, y, z cm and θ are indicated in the diagrams.

Analyse the problem	2

Investigation 1

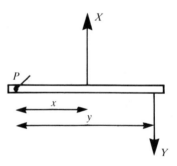

Fig. 6.8

Take moments about an axis through P:
$$Xx - Yy = 0$$
therefore
$$Xx = Yy \quad \text{or} \quad \frac{X}{Y} = \frac{y}{x}$$

Investigation 2

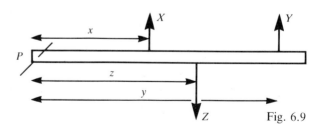

Fig. 6.9

Taking moments about an axis through P:
$$Xx + Yy - Zz = 0.$$

Investigation 3 Taking moments about as axis through P:

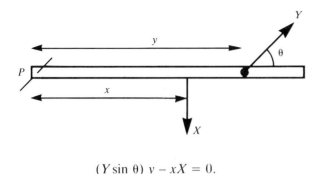

$$(Y \sin \theta) \, y - xX = 0.$$

Investigation 4 Taking moments about an axis through the centre of the rod:

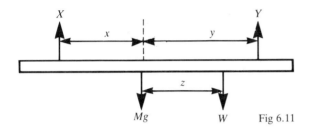

Fig 6.11

$$Xx + Wz = Yy.$$

Resolving forces vertically: $X + Y = W + Mg$.

Therefore

$$Y = \frac{W(x+z) + Mgx}{(x+y)};$$

$$X = \frac{W(y-z) + Mgy}{(x+y)}.$$

Interpret and validate 3

In each case, specific readings should be taken and substituted into the formulae.

Qualitative interpretations should also be made such as:

■ as I increase the distance y the value of Y decreases!
■ as I decrease θ the force Y has to increase.

Extensions

Determine the tensions in the chains as a window cleaner, of mass m, walks along the cradle, of mass M.

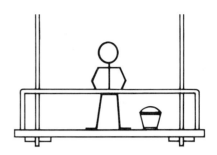

'Beam balance' provides additional problems in which the law of moments is to be applied.

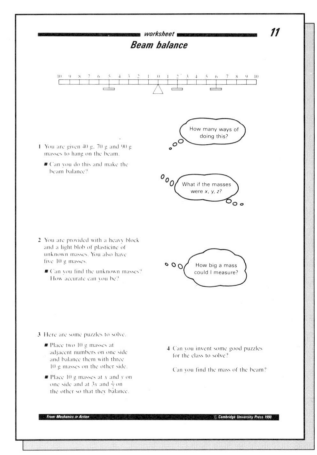

Overhang

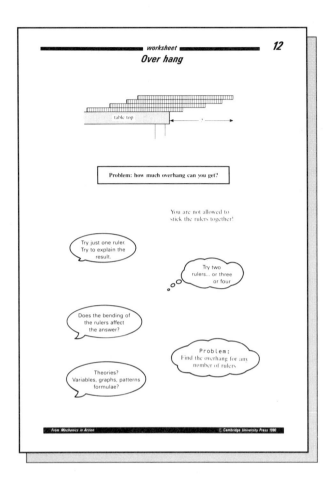

Aim

To present a challenging modelling investigation involving series and moments.

Equipment

Each group requires 4 or 5 'identical' rulers without bevels.

Plan

The problem can be presented to a group or a class via the worksheet without further explanation.

Solution

Set up model	1

A number of 1 metre rulers of weight W.

Analyse the problem	2

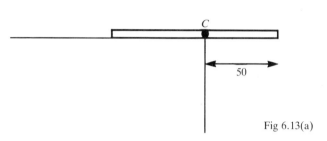

Fig 6.13(a)

1 ruler: centre of gravity C lies on the edge of the table.

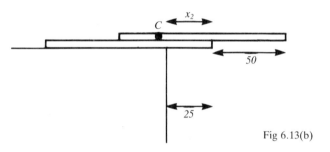

Fig 6.13(b)

2 rulers: C lies on the edge of the table, at a distance x_2 from one end of bottom ruler.

$$Wx_2 = W(50-x_2)$$
$$\text{so } x_2 = 25.$$

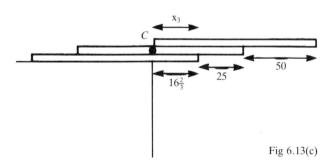

Fig 6.13(c)

3 rulers: $2Wx_3 = W(50-x_3)$
$$\text{so } x_3 = 16\tfrac{2}{3}.$$

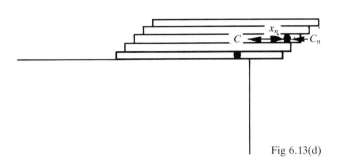

Fig 6.13(d)

With n rulers the combined mass of the top $n-1$ rulers must have its centre of gravity C_n over the end of the last, nth ruler. If x_n is the distance of C_n from the edge of the table, then the law of moments gives

$$(n-1) W x_n = W(50 - x_n)$$

$$x_n = \frac{50}{n}.$$

In general, for a ruler of length $2a$, $x_n = \dfrac{a}{n}$ and the

total overhang for N rulers is $\displaystyle\sum_{1}^{N} \frac{a}{n}$.

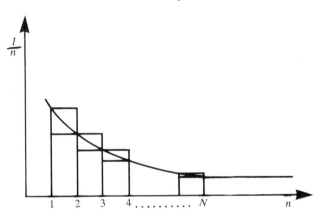

Fig 6.14

This is an interesting series. Notice that the area under the graph of $y = \dfrac{1}{x}$ satisfies

$$\sum_{2}^{N} \frac{1}{n} \le \int_{1}^{N} \frac{1}{x}\,dx \le \sum_{1}^{N-1} \frac{1}{n}.$$

It follows that

(a) $\displaystyle\sum_{1}^{N} \frac{1}{n} \le 1 + \int_{1}^{N} \frac{1}{x}\,dx = 1 + \log N$

(b) $\displaystyle\sum_{1}^{N} \frac{1}{n} \ge \int_{1}^{N} \frac{1}{x}\,dx + \frac{1}{N} = \log N + \frac{1}{N}$

and therefore

$$\log N + \frac{1}{N} \le \sum_{1}^{N} \frac{1}{n} \le 1 + \log N$$

Interpret and validate [3]

(a) The first inequality makes it clear that $\displaystyle\sum \frac{1}{n}$

has no finite limit! There is no limit to the overhang which can be achieved!

(b) Also, a prediction for the overhang of 50 rulers is $\frac{1}{2} \log 50$ ruler lengths.

Centres of mass

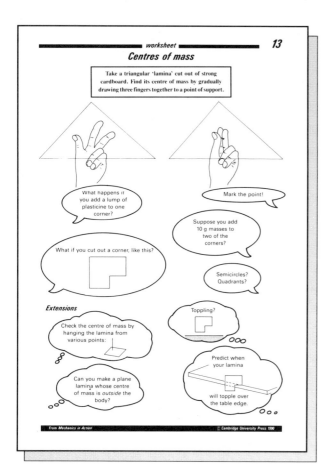

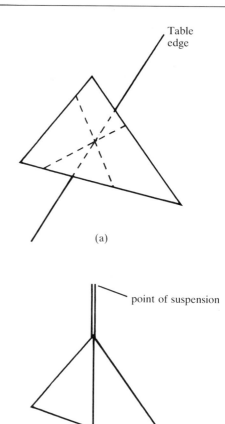

Fig. 6.15

Aims

To motivate the theory for finding the centre of mass of uniform laminas.

Equipment

Cardboard or card, scissors, plasticine, 10 g masses.

Plan

You may have different groups working with different laminas (triangles, quadrilaterals, sectors of a circle) or using different methods for finding the centre of mass.

There are 3 methods:

1 The 3 finger method – simple, quick and a little inaccurate.
2 The 'toppling edge' method – where the lamina is nudged over the edge of a table until it begins to topple and an 'edge line' is then drawn on the lamina. Clearly this line passes through the centre of mass. The method is slow but quite accurate.
3 The lamina is suspended from a point. A plumb line is then dropped from the same point and a line drawn on the lamina. The intersection of two such lines gives the centre of mass. The method is fiddly and slow but quite accurate.

This practical may be used:

(a) to obtain results before theory confirms them, for example the centre of mass of a triangle.
(b) to confirm results after the theory has been taught, for example the centre of mass of a semicircle.
(c) to provide interesting problems, for example a composite figure-problem 'for what value of x does the L-shape topple about A?'

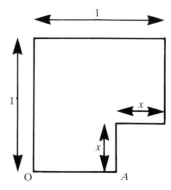

Fig. 6.16

Results can then be validated practically.

Solutions

The theory is to be found in most standard text books except possibly for the problem posed above, Fig. 6.16.

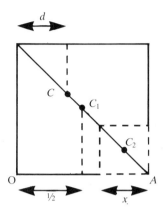

Fig. 6.17

If C, C_1, C_2 are the centres of mass of the L-shape and the squares of area 1 and x^2 respectively then the law of moments about O gives

$$(1-x^2)d + x^2(1 - \frac{x}{2}) = 1x\tfrac{1}{2}.$$

Toppling occurs when $d = 1-x$, and therefore

$$(1-x^2)(1-x) + x^2(1- \frac{x}{2}) = \frac{1}{2},$$

$$1-2x + x^3 = 0.$$

The solution of this cubic, for x, can then be found by factorisation, graph or iteration:

$$x \simeq 0.62$$

☹ Misconceptions

Intuition suggests that the centre of mass lies 'inside the lamina'. This is false as the following shapes illustrate.

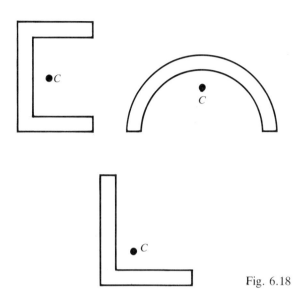

Fig. 6.18

The cable reel

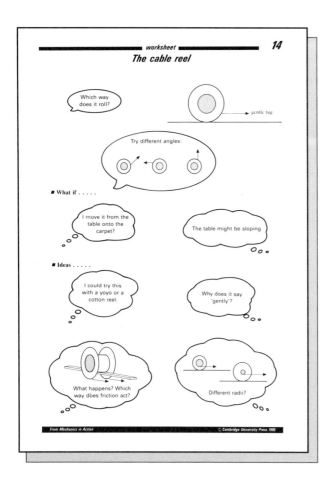

Aims

- To stimulate thought about friction.
- To identify false ideas associated with the 'laws of friction'.

Equipment

A variety of cable reels, yoyos and cotton reels. Large heavy reels work best. Light reels tend to slip on 'smooth' table tops. Yoyos and cotton reels need a guide so that they can be pulled without twisting.

Plan

A simple presentation to a class is as follows.

> 'Predict what will happen when I pull gently! Explain your prediction!'

Get everyone committed. 'Are you sure?' 'What do you think?' 'Prove it – convince me!' You may send them away to think about or analyse the problem **before** they know what happens. The actual demonstration should now provide quite a shock.

Solutions

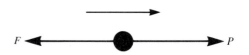

Particle model   The reel is modelled as a particle acted upon by two horizontal forces; the pull **P** and friction **F**. Students are likely to argue that the cable reel moves to the right once P is greater than F, $P > F$. This seemingly plausible argument explains the observations. Nevertheless it is inadmissible; it gives incorrect answers in other similar situations. The reel must be modelled as a rigid body.

Rigid body model   A correct solution requires a rigid body model and use of the law of moments. If T is the point where the reel is in contact with the ground, the moment of the forces about an axis through T and perpendicular to the paper is Pd in a clockwise sense. The reel will therefore turn in a clockwise sense and provided the coefficient of friction is sufficiently large, the reel will roll to the right without slipping.

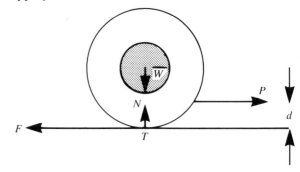

The weakness of the particle model becomes evident when other, similar situations are examined. For instance, when the 'gentle tug' is at an angle to the horizontal.

Since the tug has a component $P \cos \theta$ to the right then the particle model once again leads to the same conclusion; that the reel moves to the right once $P \cos \theta$ exceeds F!

In fact the reel rolls to the left since the moment of the pull P is now anticlockwise about T.

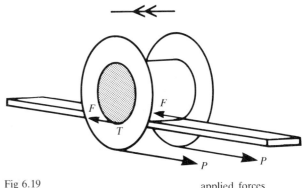

Fig 6.19 applied forces

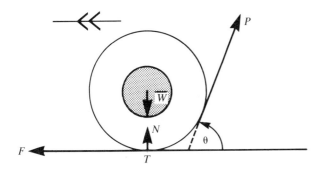

Furthermore the particle model could never explain how the reel, when placed on a track as shown, can be made to accelerate in **the opposite direction to the applied force**.

At the point of contact T, the moment of the applied forces is in an anticlockwise sense and therefore the reel moves to the left. Newton's second law describes the motion of the mass centre of the reel, and since it is towards the left then $F > P$! Friction is greater than the applied force.

These examples force us to conclude that friction can be greater than the applied force which it opposes! Friction can be in the **same direction** as the linear acceleration of the reel – as is the case with an accelerating sports car (see chapter 4, section 4.5).

The toppling tube

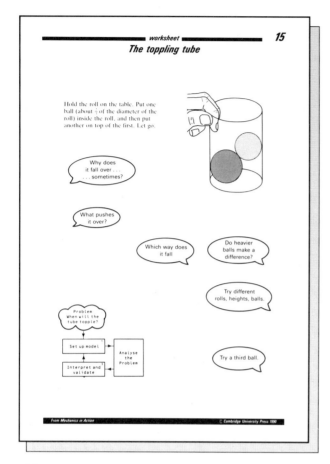

Aim

To present a challenging problem.

Equipment

Select a tube (the inside of a toilet roll perhaps) and balls whose diameters are about $\frac{2}{3}$ that of the tube. Check that the demonstration 'works' regularly for you.

Plan

Present this problem in the last 10 minutes of a lesson. Practise your act – the important thing is the element of surprise. After inserting the first ball, then a second ball (whilst holding on to the tube!), say 'What do you expect to happen when I let go?' Get everyone to justify their guess as to whether it will fall over or not!

Solutions

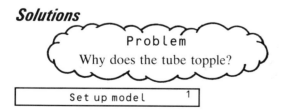

This is a three-dimensional problem which can be modelled in a two-dimensional way.

Consider first the tube standing upright on a horizontal table and in equilibrium.

The forces acting on the tube are its weight **W** and the reactions exerted by the table which are uniformly distributed around the perimeter of the base as shown.

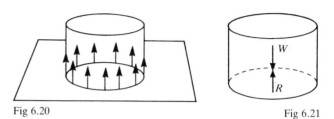

Fig 6.20   Fig 6.21

Just as we can represent the weight of the tube by a single force acting through the centre of mass, similarly the effect of these vertically directed perimeter reactions is equivalent to a single force acting through some point, which by symmetry is the centre of the base. Newton's first law gives $W - R = 0$ and the moment of forces about any point is clearly zero.

Assume now a state of equilibrium in which the balls are in the tube and their line of centres makes an angle θ with the horizontal.

Assume there is no friction force acting on the tube or the balls and let the weights and reactions be w, W, R_1, R_2, N and S as indicated. In addition there will be a distribution of reactions around the perimeter of the base which once again are equivalent to a single force through some point, which by symmetry will lie on the diameter OD. Let this force have magnitude X and let its distance from O be s.

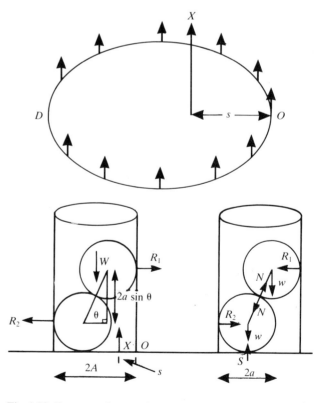

Fig 6.22 Forces acting on the tube and the balls of radii A and a respectively

From the geometry

$$2a \cos \theta = 2A - 2a$$
$$a \cos \theta = A - a. \qquad (6.1)$$

Since the balls are in equilibrium,

resolving forces horizontally

$$R_1 = N \cos \theta; \; R_2 = N \cos \theta, \qquad (6.2)$$

resolving forces vertically

$$w = N \sin \theta; \qquad (6.3)$$
$$w + N \sin \theta = S. \qquad (6.4)$$

Since the tube is in equilibrium

$$X - W = 0$$
$$R_1 = R_2. \qquad (6.5)$$

Taking moments about an axis through O

$$Xs + R_1 (a + 2a \sin \theta) - WA - R_2 a = 0. \qquad (6.6)$$

It follows from (6.5) and (6.6) that

$$2a \, R_1 \sin \theta = W(A - s)$$

and from (6.2) and (6.3) that

$$a \, R_1 \sin \theta = a \, N \cos \theta \sin \theta = aw \cos \theta$$

$$= w(A - a) \text{ from } (6.1).$$

Therefore $\quad s = A - \dfrac{2w}{W}(A - a).$

For equilibrium $s \geq 0$ and toppling is about to occur

when $s = 0$, i.e. when $\dfrac{W}{2w} = \left(1 - \dfrac{a}{A}\right).$

Toppling certainly occurs if

$$\frac{W}{2w} \leq \left(1 - \frac{a}{A}\right) \quad \text{or} \quad \frac{a}{A} \leq 1 - \frac{W}{2w}. \qquad (6.7)$$

If the weight of the tube is **small** compared to the weight of the balls, then for toppling to occur it is only necessary that $a < A$, that is, the balls fit into the tube! In general the ratio of the radius of the balls to that of the tube $\left(\dfrac{a}{A}\right)$ needs to be less than $\left(1 - \dfrac{W}{2w}\right)$ for toppling to occur. So it follows that the lighter the balls, the smaller they need to be to cause toppling. The analysis breaks down however when $a \leq \frac{A}{2}$. Therefore equation (6.7) implies that toppling cannot occur at all if $w < W$, that is, if the balls are lighter than the tube!

See Fig. 6.23

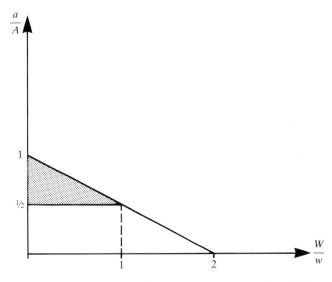

Fig 6.23 The shaded region represents cases in which toppling occurs.

☹ *Misconceptions*

The obvious misconception is for students to think of the system as 'closed'; that is, to consider the tube and the 2 balls as **one body**. Clearly a rigid body cannot topple over about a point on the edge of its base since this would imply that it gains energy as its centre of mass C rises.

Here we have an open, non-rigid system in which it is possible for the **balls** to lose height as the tube topples and its centre of mass rises. Clearly the relative masses of the balls and the tube are important.

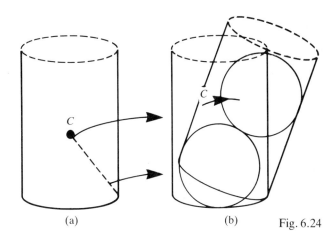

(a)   (b)   Fig. 6.24

Extensions

- Consider a tube with 3 balls.
- Consider what happens once the tube has begun to topple: can it 'rock back' into equilibrium?
- Consider 2 balls of different weights w_1 and w_2.

chapter
7
FRICTION

The practicals in this chapter all involve friction and its effect on bodies in contact. They are appropriate principally to mechanics students.

The ruler problem   can be used to introduce and motivate the study of friction. It can also be extended to provide several modelling investigations in A-level mechanics.

The 'law' of friction   provides an opportunity to validate the usual model of static friction and appreciate its limitations.

The angle of friction   introduces the concept of an angle of friction and its relationship with the coefficient of friction. It also provides a starting point for an investigation of toppling.

Least force problems (1), (2) / The ladder problem   provide modelling investigations which apply the theory of static friction to practical situations.

The ruler problem

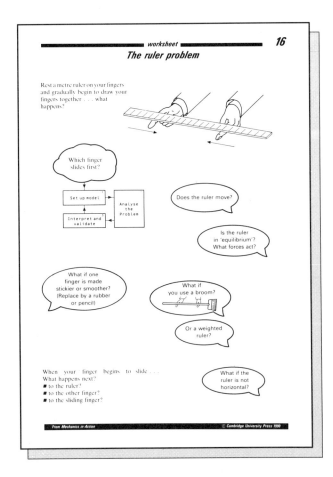

Aims

- To motivate the study of friction.
- To provide modelling problems involving a body in equilibrium subject to static friction.
- To show that static friction depends on the contact force between surfaces and has a maximum or limiting value.
- To provide extended problems to solve, involving models for static and dynamic friction.

Equipment

A number of metre rulers; a broom, Blu-tak and weights.

Plan for a 70 minute lesson and homework

Introduction (5–10 minutes)

Hold the ruler and describe the problem. Pose the question, 'What will happen when I try to draw my fingers towards each other?' At this stage **collect** all answers; get everyone committed and do not reject, qualify or criticise.

Ask for justification. Try to generate argument between students with different points of view, but avoid giving answers at this stage.

In pairs (5–10 minutes)

Give out a ruler to each pair of students. Ask them for **observations**, reasons and argument. Bring the class together to collect observations.

What happens to the ruler? Does it move? Is it in equilibrium?

Now focus on an interesting problem:

We are assuming that students have been taught the law of friction. However, if not, this introduction can serve as a motivation for the study of friction.

Group problem solving (40 minutes)

Give out worksheets and encourage students to begin to model, i.e.

(i)   draw a diagram,
(ii) insert forces,
(iii) introduce variables,
(iv) model friction.

Your questions and hints could include: Is the ruler in equilibrium? Which of Newton's laws apply? What can you say about friction? What about moments?

Your role at this stage is to intervene when necessary to prevent frustration.

Last 10 minutes

Draw the class together. Without going through the whole solution, clarify the key ideas for the weaker groups in order to help them with the homework, i.e.

(i)   the ruler is in equilibrium;
(ii) friction is the same at each finger but the reactions are different;
(iii) the law of friction, Newton's first law and the law of moments.

Groups who have solved the identified problems should be asked to examine the extensions and report.

Solutions

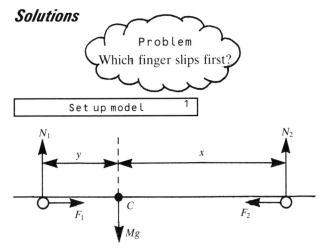

Fig 7.1 Ruler placed asymmetrically across fingers 1 and 2.

Assume:

(i)   the ruler, of mass M, is uniform, horizontal and in equilibrium. Let each finger exert a normal reaction and a frictional force as shown.

(ii) $x > y$ and the coefficient of friction is the same at each finger (i.e. each finger has the same friction characteristics!).

Analyse the problem 2

Apply Newton's first law to the ruler,

horizontally $\qquad\qquad F_1 - F_2 = 0$

vertically $\qquad\qquad N_2 + N_1 - Mg = 0.$

Take moments about an axis through C

$$N_1 y = N_2 x.$$

The law of friction states that $F_1 \leq \mu N_1$, $F_2 \leq \mu N_2$ and friction is limiting at fingers 1 and 2 if $F_1 = \mu N_1$ and $F_2 = \mu N_2$ respectively.
   The problem may now be restated as:

'At which finger does friction reach its limiting value first?'

Since $F_1 = F_2$, friction is the same at both fingers. But

$x > y$ so $N_1 > N_2$ and $\mu N_1 > \mu N_2$,
so $F_2 = \mu N_2$ when $F_1 < \mu N_1.$

Friction reaches its limiting value first at finger 2, and therefore finger 2 slips first!

Interpret and validate 3

1 The analysis confirms that the first finger to slide is the one furthest from the centre of mass, C. Intuitively, this is the finger at which the reaction is least, the finger which 'carries less of the ruler's weight', so we might naturally expect it to slip more readily there.

2 The above result enables us to deduce what happens when we use a broom or a weighted ruler. We need only locate the centre of mass and see which finger is furthest away.

Fig. 7.2

If both fingers are equidistant from C, then in principle both fingers should slide simultaneously. This is difficult to validate in practice because it is almost impossible to attain 'ideal conditions' at each finger, that is $x = y$, identical friction characteristics and a perfectly horizontal ruler.

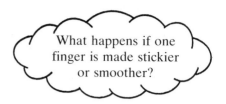
What happens if one finger is made stickier or smoother?

In this case the coefficient of friction is different at each finger and is μ_1 and μ_2 respectively.

Friction will be limiting first at finger 2 provided

$$\mu_2 N_2 < \mu_1 N_1$$

and since $\quad \dfrac{N_1}{N_2} = \dfrac{x}{y} \quad$ this is true if $\quad \dfrac{\mu_2}{\mu_1} < \dfrac{x}{y}.$

Consider the following two cases.

(i) Both fingers are equidistant from C and $x = y$. Finger 2 slides first if $\mu_2 < \mu_1.$

Validate this result by wetting the second finger with water or a thin oil or margarine. Alternatively, increase μ_1 by using a rubber instead of finger 1.
(ii) Set x to be 'slightly larger' than y; $\dfrac{x}{y} > 1.$
Then if $\mu_1 = \mu_2$ sliding should first occur at finger 2. Sliding can, however, occur first at finger 1 provided

$$\dfrac{\mu_2}{\mu_1} > \dfrac{x}{y}.$$

😕 *Misconceptions*

1 The most common misconception here is that the friction law involving F and N is an equality; $F = \mu N$. (See chapter 4, section 4.5.) This investigation clearly illustrates the significance of the inequality $F \leq \mu N$.

2 Students also tend to think that the friction forces may not be the same when slipping occurs, because the ruler may then be **moving**; this is an Aristotelian misconception. (See chapter 4, section 4.2.)

Extensions

The following problems provide extended modelling investigations for A-level course work. Both extensions are tough problems which will stretch any A-level student. See References, Davies (8) which gives data on slipping for a ruler.

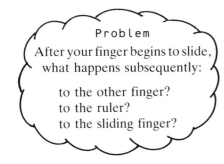

Problem
After your finger begins to slide, what happens subsequently:

to the other finger?
to the ruler?
to the sliding finger?

Set up model   1

(i) Same model as previously but now finger 2 slides and $F_2 = \mu_D N_2$ where μ_D is the coefficient of dynamic friction and $\mu_D < \mu$.

(ii) Assume that the ruler remains in equilibrium.

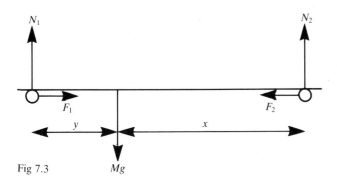

Fig 7.3

Analyse the problem   2

As before

$$F_1 - F_2 = 0$$

$$N_1 y = N_2 x$$

but now $F_2 = \mu_D N_2$ since finger 2 is slipping and $F_1 \leqslant \mu N_1$ since finger 1 is static. Therefore, $\mu N_1 \geqslant \mu_D N_2$

and $\dfrac{x}{y} = \dfrac{N_1}{N_2}$ so $\dfrac{x}{y} \geqslant \dfrac{\mu_D}{\mu}$.

When $\dfrac{x}{y} = \dfrac{\mu_D}{\mu}$ then $\dfrac{N_1}{N_2} = \dfrac{\mu_D}{\mu}$ so $F_1 = F_2 = \mu_D N_2$

$$= \mu N_1$$

and slipping occurs at finger 1.

At this instant, when $x = \dfrac{\mu_D}{\mu} y$, friction at finger 1 becomes dynamic, and there is an instant reduction in F_1 such that $F_1 = \mu_D N_1$ so $F_1 = F_2 = \mu_D N_2 = \mu N_1$. But $F_1 = F_2$

so $F_2 = \mu_D N_1 = \dfrac{\mu_D x \, N_2}{y}$.

But since $\dfrac{x}{y} = \dfrac{\mu_D}{\mu} < 1$ at this instant, $F_2 < \mu_D N_2$ and consequently finger 2 **ceases** sliding.

Interpret and validate [3]

Finger 2 slides until $x = \dfrac{\mu_D}{\mu} y$, when finger 2 reaches a point somewhat nearer C than finger 1. Note that $\dfrac{\mu_D}{\mu}$ is a number smaller than 1, and practical measurement of the slipping point gives a value of it.

At this instant, when $x = \dfrac{\mu_D}{\mu} y$, finger 1 starts to slide and finger 2 stops sliding.

Finger 1 will now slide until a new point is reached when $F_2 = \mu N_2$ again. At this instant finger 2 will start to slide but finger 1 will stop, and so on.

One can calculate the subsequent slipping points, as the fingers get closer to the centre, in terms of μ_D and μ, and initial values of x and y. More precise measurements of slipping points can be made using sharp wedges rather than fingers.

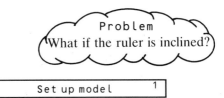

Problem
What if the ruler is inclined?

Set up model   1

Assume the ruler is at rest and inclined at an angle α to the horizontal. Otherwise the model remains the same as before.

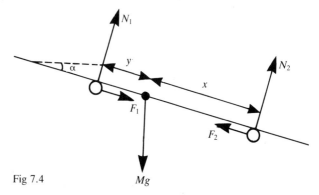

Fig 7.4

Analyse the problem   2

Resolve the forces on the ruler both vertically and horizontally

$$Mg - (N_1 + N_2) \cos \alpha - (F_2 - F_1) \sin \alpha = 0$$

$$(N_1 + N_2) \sin \alpha - (F_2 - F_1) \cos \alpha = 0.$$

Taking moments about C, $N_1 y = N_2 x$.

Divide the second equation by $N_2 \cos \alpha$

$$\left(1 + \dfrac{N_1}{N_2}\right) \tan \alpha = \dfrac{F_2}{R_2} - \dfrac{F_1}{R_1} \times \dfrac{N_1}{N_2}$$

or

$$\left(1 + \dfrac{x}{y}\right) \tan \alpha = \dfrac{F_2}{N_2} - \dfrac{F_1}{N_1} \times \dfrac{x}{y}$$

There are two cases to consider since sliding may occur first at either finger.

Case (i)   If finger 2 slides first then

$$\frac{F_2}{N_2} = \mu \quad \text{and} \quad \frac{F_1}{N_1} < \mu.$$

Therefore

$$\mu - (1 + \frac{x}{y})\tan\alpha \; < \; \mu\,\frac{x}{y}$$

$$\frac{x}{y}(\mu + \tan\alpha) \; > \; \mu - \tan\alpha$$

$$\frac{x}{y} \; > \; \frac{\tan\lambda - \tan\alpha}{\tan\lambda + \tan\alpha},$$

where $\mu = \tan\lambda$.

Case (ii) If finger 1 slides first then the inequality reverses, and we have

$$\frac{x}{y} \; < \; \frac{\tan\lambda - \tan\alpha}{\tan\lambda + \tan\alpha}$$

| Interpret and validate [3] |

The following conclusions may be drawn.

(i)   Provided $\lambda > \alpha$ then sliding will arise at finger 1 or 2 according as

$$\frac{x}{y} \genfrac{}{}{0pt}{}{<}{>} \frac{\tan\lambda - \tan\alpha}{\tan\lambda + \tan\alpha}$$

(ii) If $\lambda < \alpha$, then the assumption that the ruler is in equilibrium is false and the above model is invalid!

The 'law' of friction

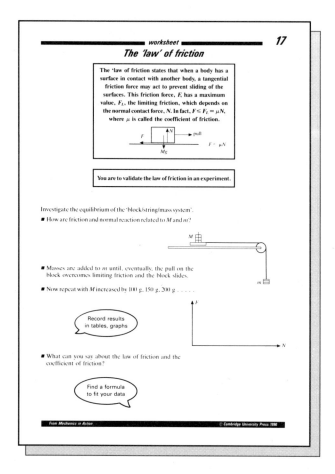

Aims

■ To establish the inequality $F \leqslant F_L = \mu N$ as a model for friction, which is both **approximate** and valid over a limited range.

Equipment

Friction plane with block and masses from the Leeds Mechanics Kit or the Physics department.

Plan

Set up the apparatus in front of the class with the block/string/mass in equilibrium. Invite the students to model the system and answer the first question on the worksheet ($F = mg$ and $N = Mg$!).

With a block of mass M_1 ($N_1 = M_1 g$), gradually increase mass m until slipping occurs. Students should appreciate that friction increases with m.

Plot the points $(N_1, F) = (M_1 g, mg)$ on a graph of F against N, circling the final value of F. Now increase the mass of the block to M_2, and repeat the experiment.

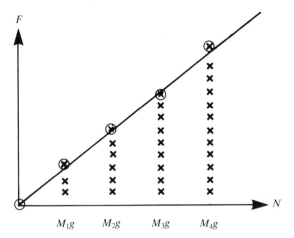

Fig. 7.5

Solution

The law of friction can be validated by means of experiment and by following the 3-stage modelling process in which the question posed is 'How is F related to N?'

Set up model   1

Assume:

(i) the pulley is smooth and the string is both light and inextensible.

(ii) the block/string/mass system is in equilibrium. Introduce forces of magnitude F, T, N, Mg, mg.

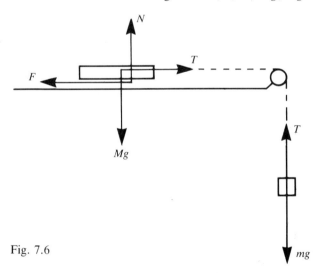

Fig. 7.6

Analyse the problem   2

Newton's first law applied to the block and mass gives, for the block

$$F - T = 0$$

$$N - Mg = 0$$

for the mass

$$T - mg = 0$$

from which we obtain

$$F = mg$$

and

$$N = Mg.$$

Interpret and validate   3

On what curve do the circled points lie? This curve is linear over some small range of N, $0 \leq N \leq N_0$ say; and it is found by drawing the best straight line through the appropriate circled points. Over this range $F_L = \mu N$ and $F \leq \mu N$ (see chapter 2).

Notes

1 The 'law' of friction is an experimental law which fits the data over a limited range only. As N increases the circled points may bend away from the straight line.

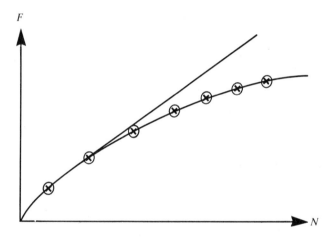

2 The surface of the plane is unlikely to be uniform; different results can usually be obtained with the block in different positions. To obtain consistency the block should therefore be positioned in the same place each time.

The angle of friction

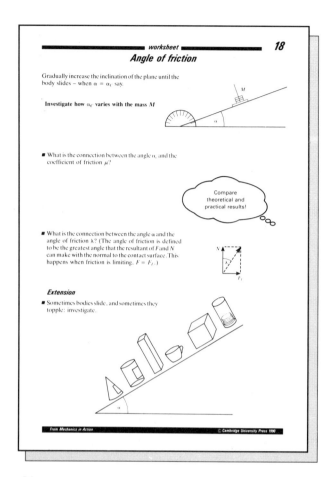

worksheet ━━━━━━ 18
Angle of friction

Gradually increase the inclination of the plane until the body slides – when $\alpha = \alpha_c$ say.

Investigate how α_c varies with the mass M

- What is the connection between the angle α_c and the coefficient of friction μ?

Compare theoretical and practical results!

- What is the connection between the angle α and the angle of friction λ? (The angle of friction is defined to be the greatest angle that the resultant of F and N can make with the normal to the contact surface. This happens when friction is limiting. $F = F_L$.)

Extension

- Sometimes bodies slide, and sometimes they topple: investigate.

From Mechanics in Action   © Cambridge University Press 1990

Aims

- To introduce the angle of friction and a simple way of finding it.
- To provide a starting point for a modelling project on toppling.

Equipment

Friction plane from the Leeds Mechanics Kit. Alternatively, use a plane leaning against a wall to which paper is attached. Draw lines on the paper and measure angles of inclination using a protractor.

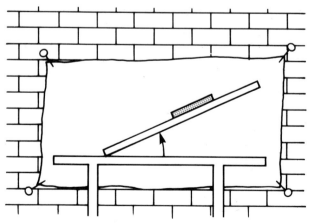

Fig. 7.7

Plan

Students can perform the experiment with blocks of different mass. Draw the best straight and horizontal line through the points (M, α) to indicate that α is independent of mass.

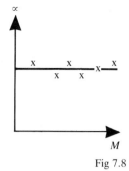

Fig 7.8

Solution

Model the block as a particle and introduce forces as shown. With the block in equilibrium and on the point of sliding, forces are resolved in two directions – parallel and perpendicular to the plane.

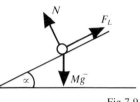

Fig 7.9

$$F_L - Mg \sin \alpha = 0;$$

$$N - Mg \cos \alpha = 0$$

$$\text{and } F_L = \mu N.$$

Therefore

$$Mg \sin \alpha = \mu Mg \cos \alpha$$

which gives

$$\tan \alpha = \mu.$$

The angle of friction is defined to be λ where

$$\tan \lambda = \frac{F_L}{N} = \mu.$$

Therefore

$$\tan \alpha = \mu = \tan \lambda$$

and

$$\alpha = \lambda.$$

When the block is on the point of slipping, the angle of inclination of the plane α is equal to the angle of friction λ. This is a simple way of finding λ!

Extensions

1 Previously the block was modelled as a particle. A rigid body of finite size, however, may well topple before it slides.

Consider a rectangular block of side $2a$, $2b$ as shown.

This will slide, as we have seen, when $\alpha = \lambda$.

As for toppling, there are 3 forces acting on the block, all of which must pass through one point P if the block is in equilibrium. (If a body is subject to 3 coplanar and non-parallel forces then any two will have a point of intersection, P. If the body is

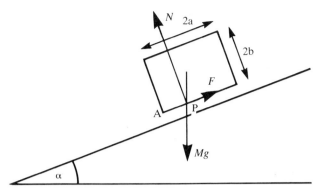

Fig. 7.10

in equilibrium the third must pass through P or there will be a non-zero moment of force about P!) As α increases P moves towards A and toppling occurs when P reaches A, that is when $\tan \alpha = \dfrac{a}{b}$.

The block will therefore slide or topple first according to whether $\lambda \gtrless \tan^{-1}\left(\dfrac{a}{b}\right)$.

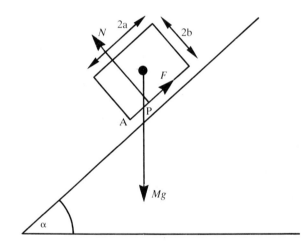

Fig. 7.11

Repeat the calculation for a cone, cylinder, cube or hemisphere.

2 A further toppling extension of interest is to investigate the toppling angle for a Smartie tube or coin holder with various coins placed inside it. This provides an interesting modelling problem: to obtain a formula for the angle, α, as a function of the number of coins, n.

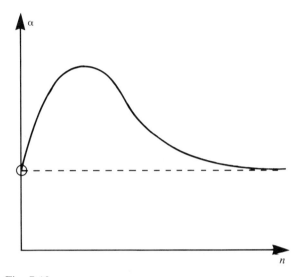

Fig. 7.12

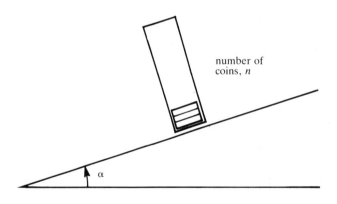

number of coins, n

Fig. 7.13

Least force problems (1)

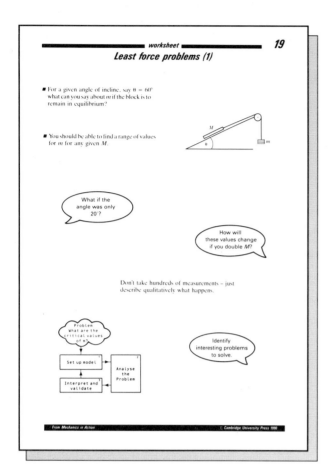

worksheet ━━━ **19**
Least force problems (1)

■ For a given angle of incline, say θ = 60°
what can you say about *m* if the block is to
remain in equilibrium?

■ You should be able to find a range of values
for *m* for any given *M*.

What if the
angle was only
20°?

How will
these values change
if you double *M*?

Don't take hundreds of measurements – just
describe qualitatively what happens.

Problem
what are the
critical values
of *m*?

Identify
interesting problems
to solve.

Set up model

Analyse
the
Problem

Interpret and
validate

From *Mechanics in Action* · © Cambridge University Press 1990

Aim

■ To present an interesting problem which illustrates
how friction opposes the tendency to slide.

Equipment

Friction plane with block and masses from the Leeds
Mechanics Kit or the Physics department.

Plan

We assume you are working with a whole class but
only have one inclined plane. Demonstrate the key
features of the situation and ask: 'Why is there a
maximum and a minimum value of *m*?' 'What does
this mean in terms of the forces acting on the block?'
Let the students then proceed with the modelling and
the analysis in groups. Finally, each group may be
invited to validate their predictions with the apparatus.

Solutions

Problems
Which values of *m* will maintain
equilibrium?
What is the minimum (and
maximum) value of *m*?

Set up model	1

Assume:

(i)   the system is in equilibrium with the block modelled as a particle and on the point of sliding **down**
the plane. For this we require $\theta > \lambda$. Forces are
as indicated with friction acting up the plane.

(ii) the string is light and the pulley is smooth.

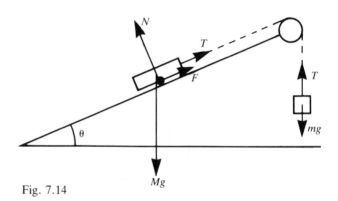

Fig. 7.14

Analyse the problem	2

Resolving forces on the mass and on the block
$$T - mg = 0$$
$$T + F - Mg \sin \theta = 0$$
$$N - Mg \cos \theta = 0$$

Therefore
$$F = Mg \sin \theta - mg \; ; \; N = Mg \cos \theta$$
and $F = \mu N$ gives
$$Mg \sin \theta - mg = \mu Mg \cos \theta$$
Therefore
$$m = M (\sin \theta - \mu \cos \theta)$$

For the maximum value of m, at which the block starts to slide up the plane, the analysis is the same with the direction of F reversed

$$F = mg - Mg \sin \theta$$

$$N = Mg \cos \theta.$$

$F = \mu N$ gives

$$mg - Mg \sin \theta = \mu Mg \cos \theta$$

and therefore
$$m = M (\sin \theta + \mu \cos \theta)$$

Interpret and validate 3

(a) The range of m which will maintain the block in equilibrium is

$$M (\sin \theta - \mu \cos \theta) < m < M (\sin \theta + \mu \cos \theta).$$

(b) As m increases through the above range the friction force changes in both magnitude and direction. In particular $F = 0$ when $m = M \sin \theta$, which follows directly from the fact that the tension T must balance the component of Mg down the plane.

(c) If M is doubled, then clearly the above range is also doubled.

(d) Note that the lower limit for m requires $\tan \theta > \mu$ in order to be positive.

Extensions

See 'The stop–go phenomenon', chapter 9.

Least force problems (2)

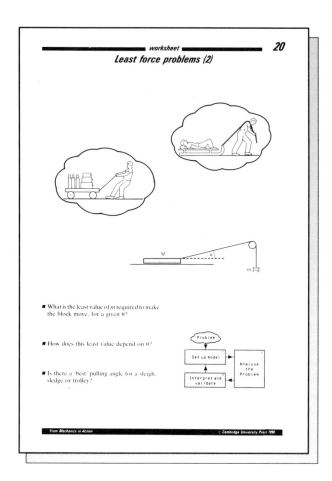

Aim

■ To present an interesting problem to solve.

Equipment

Friction plane with block and masses from the Leeds Mechanics Kit or the Physics department. Alternatively, use a **good** spring balance attached to a **heavy, rough** block.

Plan

An effective experiment is crucial in this investigation. One way is to place a rough block on a rough table with a high coefficient of friction, using tape or sandpaper if necessary to obtain $\lambda \sim 40°–60°$.

First use the friction plane with the string horizontal and find the force, T, necessary to cause sliding. Gradually increase θ and repeat. The least force to cause sliding is found to be when $\theta = \lambda$!

Solution

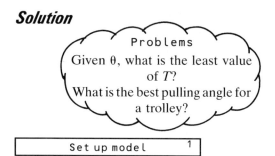

Set up model	1

Assume:

(i) the body (sleigh, sledge or trolley) can be modelled as a particle which is in equilibrium.
(ii) the pulley is smooth and the string is light and inextensible.

Introduce forces as shown in Fig. 7.15.

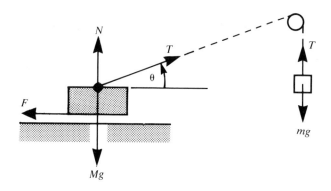

Fig. 7.15

Analyse the problem	2

Resolve forces horizontally and vertically for the block;

$$F - T\cos\theta = 0.$$

$$N + T\sin\theta - Mg = 0.$$

On the point of sliding $F = \mu N$ and therefore

$$T\cos\theta = \mu(Mg - T\sin\theta)$$

$$T = \frac{\mu Mg}{(\cos\theta + \mu\sin\theta)} = \frac{Mg\sin\lambda}{\cos(\lambda - \theta)}$$

since $\mu = \tan\lambda$.

(a) For a given value of θ, the least value of T required to cause the block to slide is $\dfrac{Mg \sin \lambda}{\cos (\lambda - \theta)}$

(see graph).

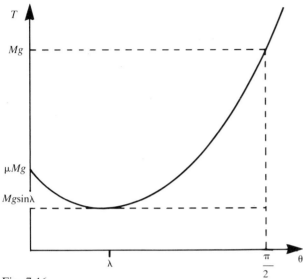

Fig. 7.16

(b) The best pulling angle θ, is clearly the one which requires the least tension T.

$T = Mg \dfrac{\sin \lambda}{\cos (\lambda - \theta)}$ is a minimum when

$\cos (\lambda - \theta) = 1$, i.e. when $\theta = \lambda$!

The best pulling angle is the angle of friction.

Misconceptions

It is often assumed that the force is least when the string is horizontal!

Extensions

1 Find the force required to push a sledge at an angle θ. What is the best pushing angle now?
For what values of θ will the sledge fail to move?

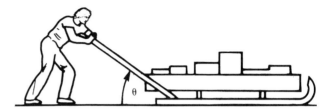

Fig. 7.17

2 Find the best pulling angle when the sledge is on an incline.

The ladder problem

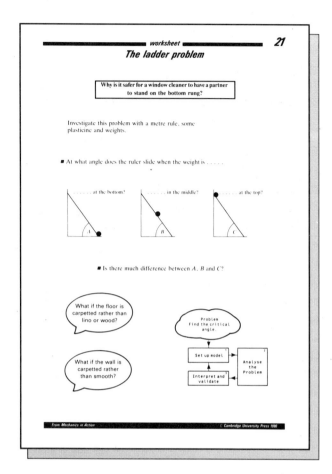

Aim

To present an interesting modelling problem which can be investigated:

- practically using simple apparatus.
- theoretically using a simple model which can be successively refined.

Equipment

Metre rule, plasticine and weights.

Plan

The practical will motivate the theory by demonstrating that the higher the window cleaner ascends the ladder, the steeper is the inclination at which the ladder slides. **Qualitative** observations are preferable to precise quantitative observations.

Solutions

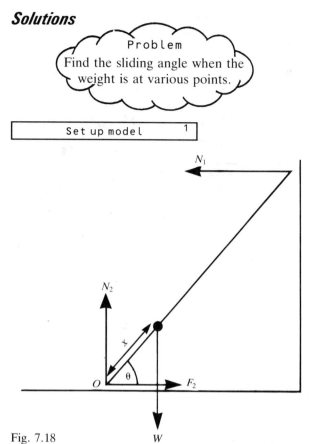

Fig. 7.18

Assume:

(i)   the ladder is light and of length $2a$.
(ii) the wall is smooth.
(iii) the ladder is on the point of sliding when a window cleaner of weight W is a distance x up the ladder.
(iv) the law of friction at the ground applies with coefficient μ.

```
Analyse the problem    2
```

Resolving forces in two directions

$$F_2 - N_1 = 0$$
$$W - N_2 = 0$$

Taking moments about O

$$2a\,N_1 \sin\theta = Wx\cos\theta$$

On the point of sliding,

$$F_2 = \mu N_2$$

therefore $N_1 = \mu N_2 = \mu W$ and

$$\tan\theta = \frac{Wx}{2aN_1}$$

giving

$$\tan\theta = \frac{x}{2a\mu} \qquad (7.1)$$

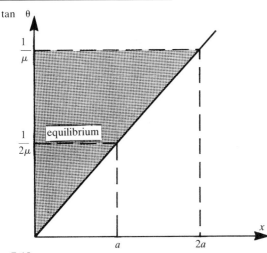

Fig. 7.19

This relationship implies that the greater x is (the higher the window cleaner ascends the ladder) the steeper is the angle at which sliding occurs. For example

when $x = 2a$; $\tan \theta = \frac{1}{\mu}$

when $x = a$; $\tan \theta = \frac{1}{2\mu}$

when $x = 0$; $\tan \theta = 0$.

Extensions

1 When a partner of weight W stands on the bottom rung of the ladder, we obtain

$$\tan \theta = \frac{x}{4a\mu}$$

As an example, consider the case in which the window cleaner is at the top of the ladder ($x = 2a$) and the ladder is inclined at an angle θ given by

$$\tan \theta = \frac{1}{\mu}$$

Then

■ without a partner – the ladder would be on the point of sliding!

■ with a partner – the ladder is perfectly safe – well above the critical angle, given by $\tan \theta = \frac{1}{2\mu}$.

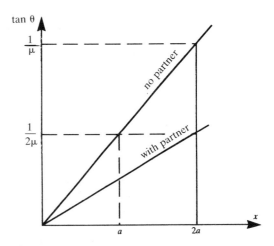

Fig. 7.20

2 If the floor is a carpet rather than lino or wood, the effect is to increase the coefficient of friction μ which reduces the right hand side of equation (7.1), thus reducing the slope of the graph. The effect is the same as having a partner on the bottom rung; it makes the ladder safer as we would expect!

3 Refine the model by including friction at the wall and also the weight of the ladder.

The formula

$$\tan \theta = \frac{(W_1 + Wx)(1 + \mu_1\mu_2) - 2\mu_2\mu_1(W + W_1)a}{2a\mu_2(W + W_1)}$$

gives the slipping angle, where W_1 is the weight of the ladder and μ_1 and μ_2 are the coefficients of friction at the wall and floor.

8

KINEMATICS AND ENERGY

This chapter contains a collection of practicals which can be used in various ways at different times in an A-level mathematics course.

The dangerous sports club problem   provides an excellent example of a modelling investigation which involves applying the theory of conservation of energy to a real problem.

Distance, speed and acceleration (1) and (2)   can be used with pure mathematics students to provide data for modelling with polynomial functions and as an introduction to calculus in the context of speed. It can also be used as a starting point for a project on rolling suitable for A-level mechanics students.

The connected masses problem   provides a practical which involves modelling with the dynamics of connected particles.

The high road and the low road   provides a practical which involves modelling with simple distance–time graphs and kinematics. It also can be extended into a modelling investigation using pure and applied mathematics including numerical methods.

Looping the loop (1)   is a practical which can be used to introduce all the concepts of mechanical energy, energy transfer, energy loss and the conservation of energy.

Projectile problem (1)   provides a practical introduction to projectiles. It can also be used simply to motivate the study of quadratic functions and their formulae.

Projectile problem (2) and (3)   can be used to validate the theory of projectiles. Various extensions are given which will provide starting points for further modelling.

The dangerous sports club problem

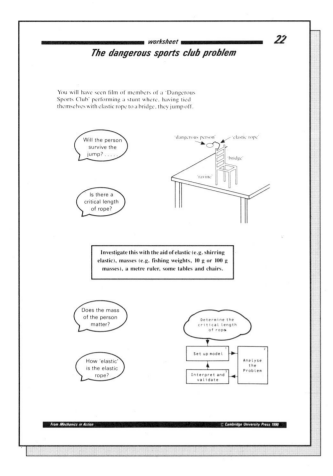

Demonstrate the problem yourself and then split the class into groups. Encourage the class to consider:

- what are the factors and variables: mass, length, type of elastic, height?
- how can we measure relevant quantities?
- what assumptions must be made?

Some groups will prefer to tackle the problem with actual measured values for elasticity, length of rope and height. Others will introduce variables immediately. Clearly a general theory is necessary to help the Dangerous Sports Club, whereas a particular result is needed to validate the theory with the given apparatus. Groups who produce a numerical solution should therefore be encouraged to introduce variables to obtain a more general result.

Solution

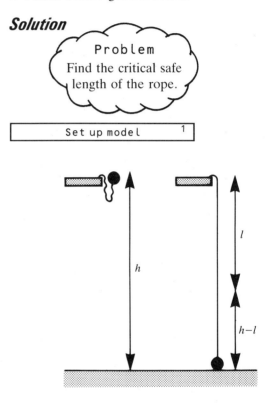

Fig 8.1

(i) Let the mass of 'diver', height of bridge, length and elasticity of rope be m, h, l and λ.
(ii) Assume Hooke's law, and no energy loss.
(iii) Assume the diver is a particle of mass m which is initially at rest, and which reaches rest again instantaneously at ground level.
(iv) Let the ground be the zero potential energy level.

Analyse the problem 2

The principle of conservation of energy gives

$$PE + KE + EE = \text{constant},$$

which implies

$$mgh = \frac{\lambda(h-l)^2}{2l}$$

Aim

- To provide a real problem to solve involving the principle of conservation of energy; elastic, potential and kinetic energy.

Equipment

The equipment is described on the worksheet. Provide various kinds of elastic: shirring and hat elastic work well. Some kinds of elastic have a plastic coating which breaks up after a few stretches: this changes its elasticity. It will therefore be necessary to get students to stretch the elastic a few times **before** beginning the experimental work. Students may need to be reminded that the elastic should not be stretched beyond its elastic limit.

Plan

70 minutes' problem solving with an applied mathematics class who have met the various forms of energy, (kinetic, potential and elastic) and the principle of conservation of energy.

Multiply both sides by $\dfrac{2l}{\lambda h^2}$ and therefore

$$\frac{2mgl}{\lambda h} = \left(\frac{l}{h} - 1\right)^2$$

or

$$\left(\frac{l}{h}\right)^2 - 2\left(\frac{l}{h}\right)\left(1 + \frac{mg}{\lambda}\right) + 1 = 0,$$

the solution to which is

$$\left(\frac{l}{h}\right) = \left(1 + \frac{mg}{\lambda}\right) \pm \sqrt{\left(1 + \frac{mg}{\lambda}\right)^2 - 1}$$

Interpret and validate [3]

1 The ratio of l to h is given as a function of the weight and elasticity.

Suppose that $\lambda = mg$, so that the rope extends to double its length when extended by the mass m (at rest). Then

$$\frac{l}{h} = 2 \pm \sqrt{3}$$

Clearly $\dfrac{l}{h} = 2 + \sqrt{3} > 1$ is not an appropriate solution.

The negative root gives a valid solution, $\dfrac{l}{h} = 2 - \sqrt{3}$.

In general,

$$\frac{l}{h} = \left(1 + \frac{mg}{\lambda}\right) - \sqrt{\left(1 + \frac{mg}{\lambda}\right)^2 - 1}$$

2 $\dfrac{l}{h}$ is a decreasing function of $\dfrac{mg}{\lambda}$. This can be sketched on a graphic calculator or function graph plotter.

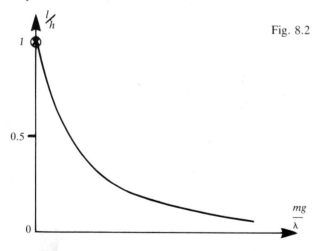

Fig. 8.2

For a given height h, the stiffer the elastic (larger λ), the longer the safe length of rope. On the other hand, heavier divers (larger mg) require shorter lengths of rope.

This can all be validated by carrying out the prac-

tical with different pieces of elastic and with different masses.

3 Students will further validate their solution experimentally by taking particular measurements for h, l, λ and m and substituting into the formula.

The model can then be used to estimate lengths for various weights of diver for a typical rope as follows.

Say $\lambda = 800$ N, a bridge has height 30 m, and the divers vary in mass from 50 to 80 kg.

Taking $g = 10$ gives critical lengths as follows. For 50 kg,

$$l_1 = 30\left[\left(1 + \tfrac{5}{8}\right) - \sqrt{\left(1 + \tfrac{5}{8}\right)^2 - 1}\right] = 10.3 \text{ m}$$

and for 80 kg,

$$l_2 = 30 \times (2 - \sqrt{3}) = 8.04 \text{ m}$$

You may ask: 'How much further allowance would you make before **you** would be prepared to do this jump?' Clearly you need to take account of the fact that you are not a particle, and you are attached to the rope by some kind of harness. Furthermore you may want a safety margin!

Extensions

The variety and depth of extensions to this problem recommend it as a piece of open-ended A-level coursework.

1 The excitement of the dive depends on several factors:

   (i)   the free-fall time of the diver,
   (ii) the height of the bridge,
   (iii) the large acceleration of the diver.

Further investigation of the problem could involve studying these factors.

2 The existing model implies that the diver carries on bouncing up and down forever. A refined model is required to take account of energy losses in the rope. 'How many bounces does the diver in fact make?' 'Can you estimate the energy losses in your experiment?'

3 A ready reckoner or rule of thumb could be made for a particular rope which allows different divers to calculate critical lengths for various bridges.

4 Real ropes do not obey Hooke's law precisely. When the elastic limit is reached the sport becomes **extremely** dangerous. It should be possible to calculate the greatest height of bridge for any given rope so that a diver will not exceed this limit.

5 Examine other alternative extension laws for the rope, such as

$$T(x) = T(0) + \frac{\lambda x}{l}, \text{ (pretensioning model)}$$

or

$$T(x) = T(0) + T_1 x + T_2 x^2 + \ldots$$

Distance, speed and acceleration (1) and (2)

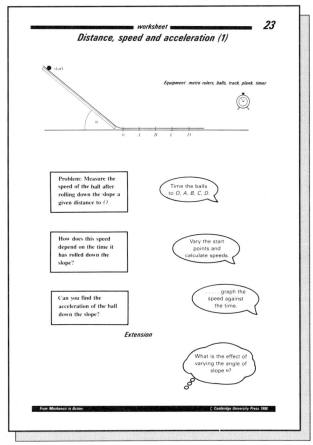

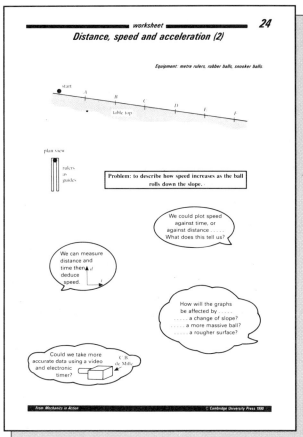

Aims

- To provide a situation to model empirically using functions.
- To provide a situation to model using either the kinematics or the dynamics of motion under constant acceleration or both.
- To provide a starting point for extended project work on the rolling of rigid bodies (trolleys, etc.)

Equipment

Metre rulers, stop watches and various snooker balls or rubber balls. Track is provided by Unilab in the Practical Mathematics Kit.

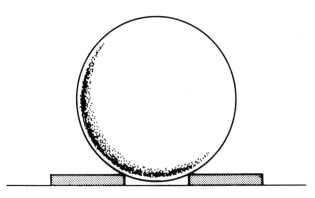

Fig. 8.3

The idea is to roll the ball down a track without it bumping, sliding or otherwise incurring energy losses. Balls without ridges roll well down some tracks. Rulers used as tracks for a rolling ball work very well (see Fig. 8.3). A snooker or pool ball rolling down a flat and inclined table will also work well. It is important that the angles of incline are **small**, so that the motion takes up to 3 or 4 seconds and can be accurately timed.

Plans

A whole class can work together to collect data, but students enjoy getting their hands on this in small groups and tackling the practical problems themselves.

70 minutes can be spent in groups collecting 'good' data and beginning to process it. By 'good' data we mean that the results obtained:

(a) are **consistent**, i.e. do not vary much from one trial to another.
(b) are **accurate**, i.e. that different people's times for the same run do not vary too much.

If the whole class conducts an experiment under your guidance, then this work can take as little as 15 minutes.

Solution A

The mathematics expected is of two kinds. In this section we describe how the experimental data gives distance–time and speed–time graphs which model the motion. The concept of gradient is applied and interpreted. This provides an informal introduction to calculus through modelling motion.

(i) From sheet 1, plot the data on a graph and estimate the gradient of the graph at the origin to obtain the ball's speed at O (Fig. 8.4).

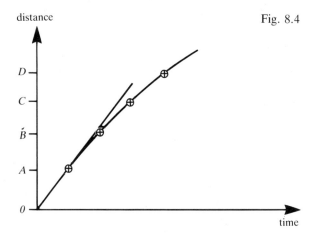

Fig. 8.4

Then take these speeds (for different release positions up the incline) and plot them on a graph against the time of descent (Fig. 8.5).

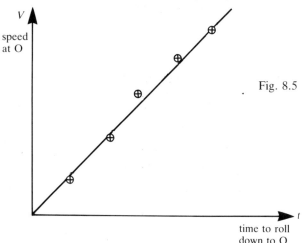

Fig. 8.5

A formula should now be found which fits the data

$$V = at$$

for some a.

(ii) From sheet 2, distance–time data can be modelled using a function graph plotter or graphic calculator (see Fig. 8.6).

A formula should now be found to fit the data

$$d = kt^2$$

for some k.

If the graph plotter has the facility to plot a first derivative (such as FGP), then a formula for this can also be found which fits the graph

$$v = at$$

for some a.

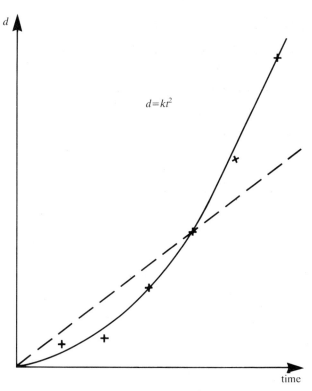

Fig. 8.6 Distant–time data gives $d=kt^2$. The first derivative (dotted) is linear, showing constant acceleration.

Students will appreciate that the ball has a constant acceleration. You will be able to relate the various formulae for the standard results of calculus

$$d = kt^2 \quad \text{or} \quad \tfrac{1}{2}at^2 \quad \text{or} \quad \tfrac{1}{2}vt$$
$$v = 2kt \quad \text{or} \quad at \quad \text{where}$$
$$a = 2k$$

and by transformations of formulae

$$v^2 = 2ad \quad \text{or} \quad v = \sqrt{2ad}.$$

Different groups using different slopes will obtain different accelerations. Alternatively, different balls can be given to different groups to find their accelerations on a given slope.

This will motivate the study of the dynamics of one-dimensional motion which follows in 'solutions B'. Of course, if you choose to use this practical with a class which has already met the theory, then you will expect them to use this in **predicting** the above graphs. The experimental work will then be reduced to taking measurements and validating the theory.

Solutions B

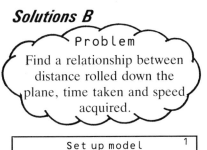

Problem

Find a relationship between distance rolled down the plane, time taken and speed acquired.

| Set up model | 1 |

Assume the ball is a particle sliding down a smooth plane inclined at an angle θ.

Let d and v be the distance and speed of the ball after time t.

| Analyse the problem | 2 |

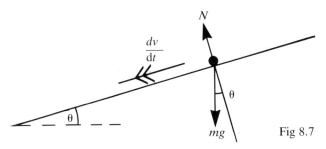

Fig 8.7

Apply Newton's second law to the motion of the particle parallel to the plane.

$$mg \sin \theta = m \frac{dv}{dt}$$

which implies acceleration,

$$\frac{dv}{dt} = g \sin \theta,$$

$$v = (g \sin \theta) t$$

and

$$d = \tfrac{1}{2} (g \sin \theta) t^2.$$

Furthermore,

$$v^2 = 2gd \sin \theta.$$

| Interpret and validate | 3 |

1 The acceleration is constant and increases as the sine of the angle of slope. It is proportional to the height of one end of the plane.

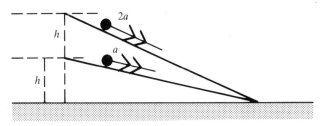

Fig. 8.8 To double the acceleration double the height of the slope.

2 Times and distances can be calculated. For example $g = 10$, $d = 1$, $\theta = 30°$, implies $t = 0.6$, so it is predicted that the ball will take 0.6 seconds to roll down 1 metre.

Such predictions will be **inaccurate**, since our model above takes no account of rolling energy!

3 The equation $\qquad v^2 = 2gd \sin \theta$

which implies $\quad \tfrac{1}{2}mv^2 = mg (d \sin \theta)$

is an energy equation for a particle, and provides a new way of thinking about the problem:

$$\text{KE gained} = \text{PE lost.}$$

This motivates a refined model for the situation and extended investigations of other rolling bodies.

Extensions

1 Consider a refined model which incorporates the rotational energy of a rolling ball.

$$\tfrac{1}{2} I\omega^2 = \tfrac{1}{2} (\tfrac{2}{5} ma^2) \omega^2 = \tfrac{1}{5} mv^2.$$

(See chapter 1, section 1.2(f).)

$$\text{KE gained} = \tfrac{1}{2} mv^2 + \tfrac{1}{5} mv^2 = mgd \sin \theta.$$

So

$$v^2 = \frac{10}{7} gd \sin \theta.$$

2 If there remain substantial discrepancies, consider refining the model to include friction. Energy loss = rolling friction × distance travelled.

3 Model the rolling of various bodies which have rolling wheels. Note that the chassis does not roll, for example, on a trolley (see Fig. 8.9).

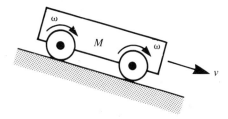

Fig. 8.9 KE of chassis $= \tfrac{1}{2} Mv^2$

$\qquad$ KE of wheel $= \tfrac{1}{2} mv^2 + \tfrac{1}{2} I\omega^2.$

Incorporate the rotational energy of the wheels and energy loss due to friction as necessary to provide a valid model for a **particular** trolley that you have experimented with.

The connected masses problem

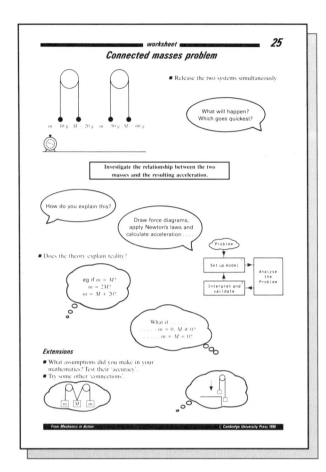

Aims

- To provide a practical problem to solve using the theory of 'connected particles'.
- To illustrate the importance of the assumptions that pulleys are 'smooth' and strings 'light'.

Equipment

'Smooth' pulleys (which can be hand-held) as provided by Unilab in the Leeds Mechanics Kit and in the Practical Mathematics Kit. String, 10 gram masses and holders. 1 metre rulers and stop watches.

Plan

Present this to the class as a whole or in groups. Rather than taking lots of accurate measurements, encourage qualitative observations such as 'for a fixed difference between m and M, the larger the total mass $(m+M)$ the smaller the acceleration'.

It is assumed that students have covered kinematics and Newton's second law. Acceleration can be measured more accurately if there is a large distance for the mass to fall to the floor.

Solutions

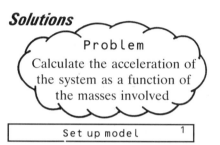

Problem

Calculate the acceleration of the system as a function of the masses involved

Set up model	1

Assume the string is light and inextensible, the pulley is smooth and the masses are particles of mass m and M. Let the acceleration be a.

The tension is a constant, T, throughout the string.

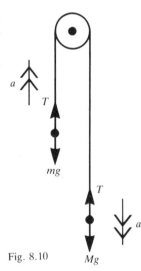

Fig. 8.10

Analyse the problem	2

Newton's second law gives

$$T - mg = ma$$
$$Mg - T = Ma$$

which implies $\quad a = \left(\dfrac{M-m}{M+m}\right)g.$

For a given fixed **difference** in mass, the system with the least total mass will accelerate more quickly. This can be validated.

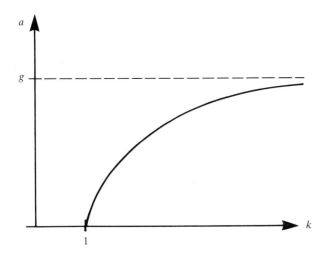

Fig. 8.11

1 The acceleration depends on the **ratio** of the masses rather than their difference. To make this explicit, set $M = km$. Then

$$a = \left(\frac{k-1}{k+1} \right) g$$

or

$$a = \left(1 - \frac{2}{k+1} \right) g.$$

2 A graph of a as a function of k reveals two obvious interpretations.

   (i) When the masses are equal, $k = 1$, $a = 0$, no acceleration takes place. Clearly for $k < 1$, $a < 0$ and $k > 1$, $a > 0$, so the direction of acceleration depends on which mass is the larger.
   (ii) When $k \to \infty$, $a \to g$. So, if $m = 0$, M will have acceleration g, as if it were disconnected.

3 The result can be used to provide a **numerical** interpretation; for example for $m = 10$, $M = 20$, $a = \frac{1}{3}g \sim 3\frac{1}{3}\,\mathrm{m\,s^{-2}}$. This prediction can be tested by timing the fall of the mass to the floor and using $d = \frac{1}{2}at^2$ to measure a.

The high road and the low road

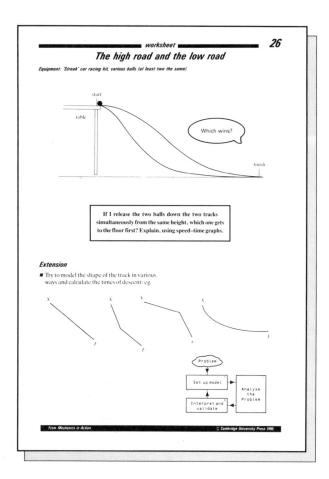

Aims

- To provide a problem to solve using kinematics and speed–time graphs.
- To provide an extended modelling project involving the calculation of times of descent.

Equipment

'Streak' car racing kit as provided by Unilab in the Leeds Mechanics Kit. Various balls (at least two the same).

Plan

The problem may be introduced in the last 10 minutes of a lesson to provide a stimulating problem to solve for homework. Alternatively, you may need to be on hand to help students discuss speed–time graphs.

As an introduction to the problem you should simply ask the students to guess which ball gets to the finish first. Get everyone to make a prediction. A common response is to say 'they both arrive together', which may be justified by:

(a) conservation of energy arguments: they both fall through the same height, or

(b) symmetry arguments: the first goes faster along AB but the second makes up for this by going faster along CD.

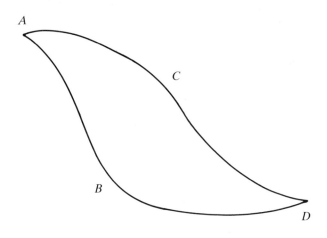

It is worth trying it out in practice before the class becomes convinced **theoretically** that ABD is the faster of the two routes, because this provides a pleasing feeling of surprise and a need to satisfy oneself of the theory behind the result. Encourage the use of speed–time graphs in their explanations.

Solution

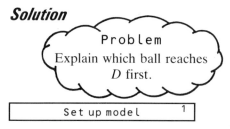

A simple model of the situation is to assume that the paths are straight sections of wire of length l, and the rolling balls are smooth beads on the wire, so there is no friction or loss of contact. The accelerations of the beads along the wires are then $g\cos\theta$ and $g\sin\theta$.

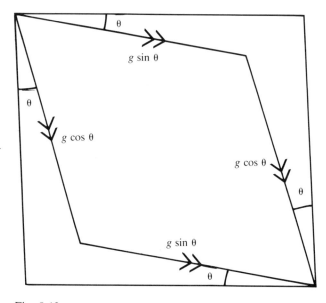

Fig. 8.12

Speed–time graphs of the two routes can be drawn using the fact that the gradients of the sections are $g \cos \theta$ and $g \sin \theta$ and the areas under the two sections are both equal to l.

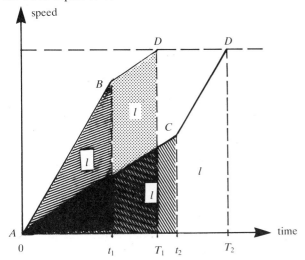

Fig. 8.13 The total times to descend to D are T_1 and T_2. The split times are t_1 and t_2 for routes ABD and ACD respectively.

Finally, since there is no energy loss the velocities at D are equal, so that end points of the two graphs have the same height.

1 The graph shows that $T_1 < T_2$. Observe that $t_1 < t_2$ and $T_1 - t_1 < T_2 - t_2$, i.e. each part of the journey takes less time for the low road than the high road.
2 The graph also implies that $0 < t_1 < T_1 < t_2 < T_2$,

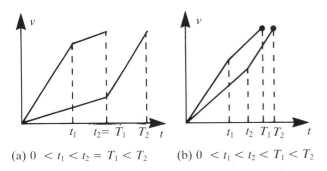

(a) $0 < t_1 < t_2 = T_1 < T_2$ (b) $0 < t_1 < t_2 < T_1 < T_2$

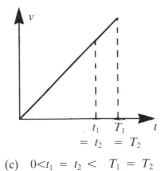

(c) $0 < t_1 = t_2 < T_1 = T_2$

Fig 8.14

which may **not** be the case. In other words the ball on ACD may arrive at C before the ball ABD finishes its run. It all depends on the value of θ and therefore on the slopes of the two graphs. As θ approaches 45°, the speed–time graphs approach a straight line as shown in Fig. 8.14.
3 Graphical interpretation leads to the conclusion that T_1 is least when $\theta = 90°$.

😞 *Misconception*

Both balls arrive together!

Extensions

Calculating the times of descent for different shapes of track may prove very demanding. It may involve the creation and testing of some numerical algorithms with the aid of a computer (either writing short programs or using a dynamic modelling system). Fig. 8.15 gives some analytical solutions, assuming the ball is a **smooth particle**.

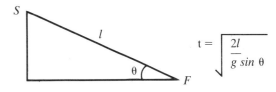

Fig 8.15(a) Particle sliding down a smooth wire from start S to finish F

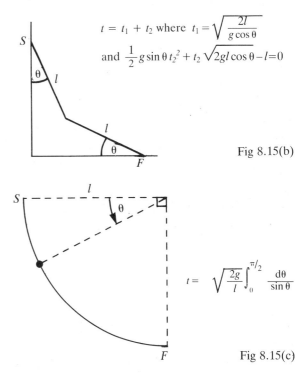

$t = t_1 + t_2$ where $t_1 = \sqrt{\dfrac{2l}{g \cos \theta}}$

and $\dfrac{1}{2} g \sin \theta\, t_2^2 + t_2 \sqrt{2gl \cos \theta} - l = 0$

Fig 8.15(b)

$t = \sqrt{\dfrac{2g}{l}} \displaystyle\int_0^{\pi/2} \dfrac{d\theta}{\sin \theta}$

Fig 8.15(c)

Students might like to read about the classic 'Brachistachrome Problem' tackled by the Bernoulli brothers, and independently by Leibniz and Newton.
The problem is 'what shape of path gives rise to the shortest journey time from A to D?'

Looping the loop (1)

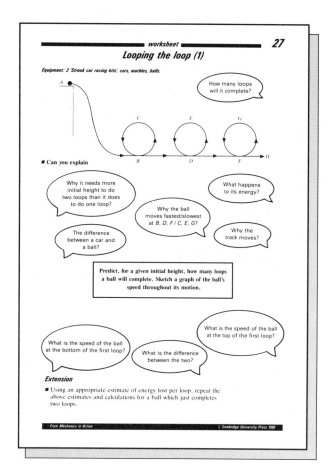

Aims

■ To provide a problem to solve using the principle of conservation of energy and the concepts of energy transfer and energy loss.
■ To demonstrate the significance of the assumption that energy is conserved in different situations.

Equipment

Two 'Streak' car racing kits as provided by Unilab in the Leeds Mechanics Kit. Various cars, marbles and balls.

Plan

The worksheet and some 'Streak car racing kits' can be given to groups of students. The key questions are:

■ what happens to the energy of the car or ball?
■ where does the lost energy go?
■ how do kinetic energy and potential energy change throughout the motion?

The groups should be able to find critical heights for n loops, which gradually increase because of the energy loss between loops. At this point, you may gather the class to discuss their results. Concentrate on the con-

cepts of energy transfer and energy loss. Ask for the sketch graphs from different groups and compare them on the board.

The class should be ready for some **quantitative** analysis. They may already know that potential energy $= mgh$ and kinetic energy $= \frac{1}{2}mv^2$.

Ask them to **calculate** the speed of the ball at various points:

(i) assuming no energy loss.
(ii) assuming losses due to friction using the data they collected.

Solutions

The difference between the critical heights for one loop and two loops is explained by the energy lost due to friction between the two loops. A reasonable, but not precise, model of this is that it is proportional to the distance moved. For a **ball**, the friction losses are small if the approach slope is gentle and if the track is braced. For the **car**, energy loss is relatively large, and therefore needs to be built into any model.

Graphs of speed against time should look like those in Fig. 8.16.

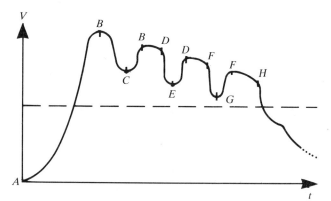

(a) A ball which completes 3 loops

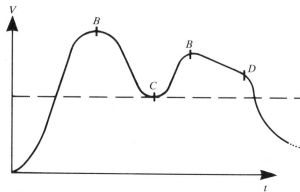

(b) A car which completes one loop

Fig. 8.16

The dotted line represents the critical speed for a particle at C, E or G, if it is not to fall off the track. Calculations of potential and kinetic energy of the ball or car will depend on the parameters involved.

Projectile problem (1)

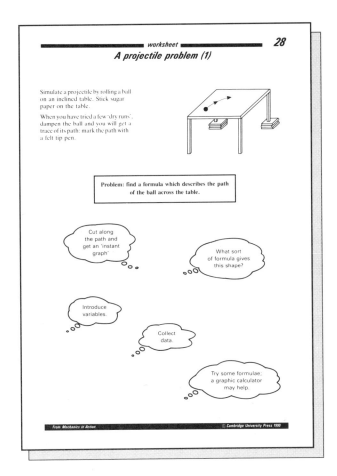

Aims

- To provide a practical simulation of a projectile.
- To provide data to model with quadratic functions on a graph plotter.
- To provide a projectile problem to investigate with Newton's laws.

Equipment

Sugar paper and a flat inclined table. A squash ball, snooker ball or pool ball.

Plan

Either give the worksheets and equipment to different groups or demonstrate from the front and select different curves for different groups to model. The graph can be cut out of the paper and stuck on to squared paper to get numerical data more easily.

The data should be entered onto a function graph plotter or graphic calculator. Functions will be found by trial and improvement. This provides a rich experience of quadratic functions and their graphs. If the class study A-level mechanics, they should be encouraged to model the situation using projectile theory. One approach is given in solution 2.

Solution 1

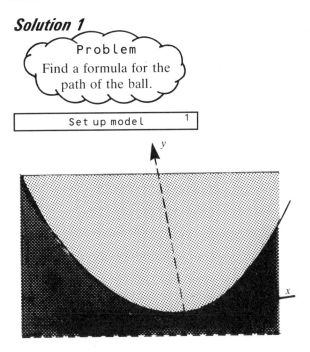

Introduce x and y axes using the symmetry of the curve as shown in Fig. 8.17.

Choose a scale for x and y, label the axes and record points on the curve in a table.

> Analyse the problem   2

The problem now is to find a function which fits the data. Try $y = ax^2$ and vary a until the curve fits the points plotted. If no function graph plotter is available, plot y against x^2 and read off the gradient.

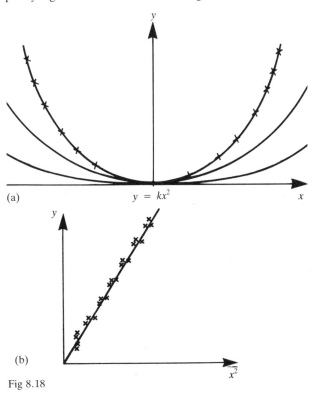

Fig 8.18

The formula gives a description of one path. However, it will not be possible to make quantitative predictions about **new** paths: this will require an approach using Newton's laws.

Solution 2

Set up model   1

Assume the ball is a smooth particle projected up the table with speed V and at an angle θ to the horizontal. Let the angle of greatest slope be α to the horizontal, and let X and Y be the horizontal and 'vertical' axes on the table.

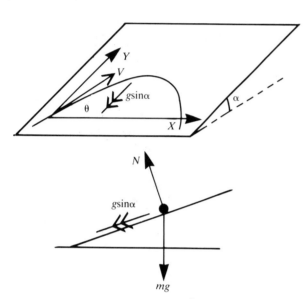

Fig. 8.19

Analyse the problem   2

The particle accelerates down the slope with acceleration $g \sin \alpha$. The usual integration with respect to time gives

$$Y = X \tan \theta - \frac{g \sin \alpha \, X^2}{2V^2 \cos^2 \theta}$$

1 Entering this formula into a graph plotter, various shapes of paths can be found which you should be able to reproduce on the table top by varying θ and V.

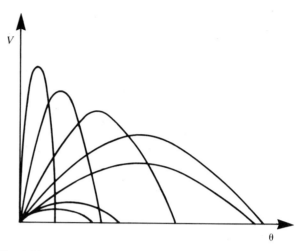

Fig. 8.20

2 For any parabola the coordinates of the vertex are:

$$(X, Y) = \left(\frac{V^2 \sin 2\theta}{2g \sin \alpha}, \quad \frac{V^2 \sin^2\theta}{2g \sin \alpha} \right)$$

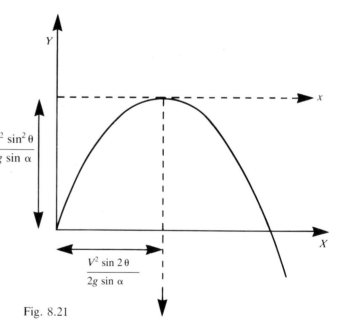

Fig. 8.21

Fig. 8.21 suggests the transformation

$$X = \frac{V^2 \sin 2\theta}{2g \sin \alpha} + x$$

$$Y = \frac{V^2 \sin^2 \theta}{2g \sin \alpha} - y$$

to obtain the parabola in standard form

and so

$$\frac{V^2 \sin^2 \theta}{2g \sin \alpha} - y = \left(\frac{V^2 \sin 2\theta}{2g \sin \alpha} + x \right) \tan \theta$$

$$- \frac{g \sin \alpha}{2V^2 \cos^2 \theta} \left(\frac{V^2 \sin 2\theta}{2g \sin \alpha} + x \right)^2$$

which eventually reduces to

$$y = \frac{g \sin \alpha}{2V^2 \cos^2 \theta} x^2.$$

This should be compared with the result $y = kx^2$ obtained in solution 1.

Comments

It is possible to control the speed of projection by releasing the ball from a chute made from a piece of kitchen roll tube and the corner of a box such as a cereal packet.

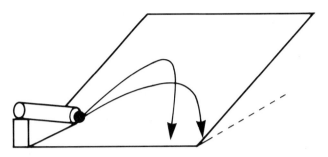

Fig. 8.22

However, care must be taken to avoid steep angles of the table or chute; these may lead to significant impacts occurring when the ball exits the chute and strikes the table.

Note also that the height of the end of the tube where the ball is placed must be the same each time for the projectile velocity to be the same. Care must also be taken to try to get the same initial 'point of projection' if comparisons are to be made.

Projectile problem (2) and (3)

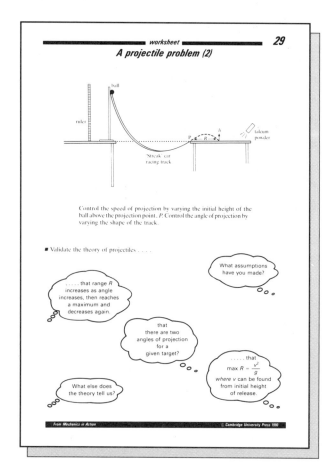

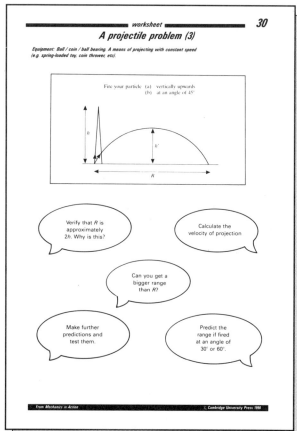

Aims

- To interpret and validate the theory of projectiles in practical situations.
- To appreciate the importance of the assumptions on which the theory is based.

Equipment

Ruler, ball, talcum powder, 'Streak' car racing track as provided by Unilab in the Leeds Mechanics Kit. Ball, coin, ball bearing. A spring-loaded toy, such as a coin thrower, which projects the coin or ball with constant speed.

Plan

30 minute activity for A-level mechanics students who have studied the theory of projectiles

As a whole class or in small groups set up the apparatus and tackle the questions on the worksheets.
Key questions are as follows,

- What are you **assuming** when you apply projectile theory?
- Are these assumptions valid?
- How accurate should you expect the predictions of the theory to be?

Do **not** expect that all predictions will be validated, it is important only that the students come to appreciate that in some cases the assumptions are too severe.

NB The variation in the projectile's release velocity can be estimated simply by observing the change in h over a number of shots.

Solutions

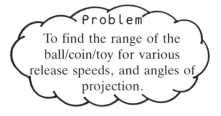

The vital assumption is that the projectile speed, V, and angle, θ, are consistent, independent and measurable. R, h and h_1 are the ranges and heights shown in the diagram on the worksheet.

The usual analysis of the motion of a particle projectile in two dimensions subject to no air resistance and constant gravity g, gives

$$R = \frac{V^2}{g}\sin 2\theta$$

$$h_1 = \frac{V^2}{2g}\sin^2\theta$$

$$h = \frac{V^2}{2g}.$$

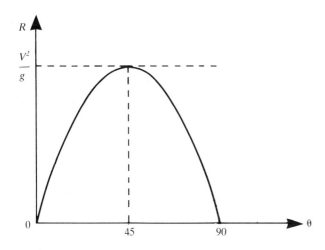

Fig 8.23

The range varies with the angle of projection for fixed velocity, as in Fig. 8.23. This may only be validated if the velocity of projection really is independent of the angle. Changing the angle of projection can either:

(i) change the slope of the approach track and so change the energy loss of the ball sliding down the track.
(ii) change the performance of the spring-loaded gun or toy.

Extensions

Much satisfaction can be gained from making predictions based on theory and then testing them out. For instance, one can calculate the angle α to hit a given target or to get the shot into a basket.

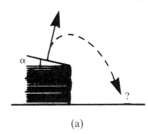
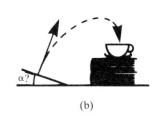

(a)   (b)

Fig 8.24

9
CIRCULAR MOTION

This chapter is divided into two parts, introductory practicals and modelling investigations.

In part A there are two practicals introducing the kinematics and dynamics of circular motion.

Pennies on a turntable   provides a practical context in which the kinematics can be taught, i.e. **v** is tangential of magnitude $v = r\omega$ and **a** is centripetal of magnitude $a = r\omega^2$.

Conical pendulum (1)   provides a practical context in which to introduce the dynamics of circular motion and the need for a resultant force towards the centre of the circle so that $\mathbf{F} = m\mathbf{a}$.

Both 'Pennies on a turntable' and 'Conical pendulum (1)' also lead naturally to modelling investigations in which the principles of circular motion are applied to practical problems as in part B.

Part B consists of eight modelling investigations involving motion in both horizontal and vertical circles.

- **Banking**
- **The rotor**
- **Conical pendulum (2)**
- **Chairoplanes**
- **Looping the loop (2)**
- **Wind-up**
- **A stop–go phenomenon**
- **Cake tin**

The notes provide occasional guidance for plans and equipment. Lessons can be organised in a number of different ways depending on the skill and experience of the students and on how much equipment you have available. In each case it is assumed that the students have studied the theory and are now being asked to apply it in a practical situation.

Plan 1   You can introduce the practical to the whole class, involving them in the identification of problems and discussion of the model-building assumptions. This may lead to individuals, pairs and groups solving the problems and then interpreting and validating their results.

Plan 2   Worksheets and equipment can be given to groups of students and you can 'stand back' and act as a consultant on request.

In either case each practical leads to modelling investigations which can be followed up by individuals or groups with an extended piece of work or project. The 'Extensions' section in each case lists a number of suggestions for these. The mathematical notes only provide examples of possible solutions to some of the problems. They are not complete or definitive.

Pennies on a turntable

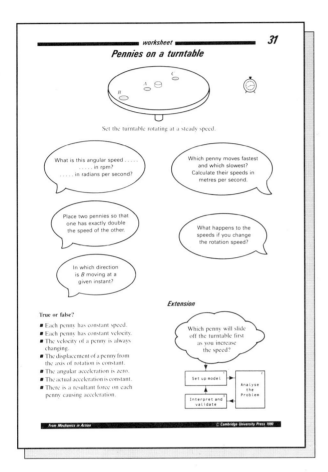

Aims

■ To focus discussion on the key concepts involved in circular motion.
■ To provide real problems to investigate and model with the theory of circular motion.

Equipment

A turntable with variable speed motor, as provided by Unilab in the Leeds Mechanics Kit. Stopwatches.

Plan

An introduction to circular motion for an A-level mechanics class.

It is assumed that the class have covered radian measure and are familiar with the basic definitions in kinematics.

If you only have one motor unit then a discussion with the whole class is appropriate. You should keep the worksheet for yourself in this case and pose a series of questions or problems to solve.

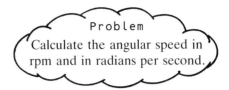

Problem
Calculate the angular speed in rpm and in radians per second.

This will test their knowledge of radians by applying it to a practical situation.

Set the turntable rotating and ask the class to determine its speed in rpm and then in radians per second. This involves counting the number of revolutions in one minute, say, dividing by 60 and multiplying by 2π

$$(\text{radians per second}) = \frac{2\pi \times (\text{no. of revs in one minute})}{60}$$

Let T seconds be the time for one revolution.

Then

$$T = \frac{60}{(\text{no. of revs in one minute})}$$

and we obtain the equation

$$\omega = \frac{2\pi}{T}. \qquad (9.1)$$

Note that ω, the number of radians per second, can also be regarded from (9.1) as a **frequency**, the number of revolutions in a time of 2π seconds.

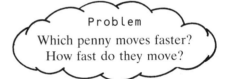

Problem
Which penny moves faster?
How fast do they move?

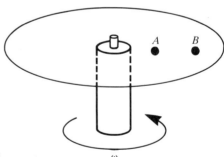

Fig. 9.1

Place two pennies at different distances, r_1 and r_2, from the axis and ask the class 'Which goes faster?'. Clarify the distinction between angular speed, ω, and 'actual' speed. Some students will know that speed v is given by $v = r\omega$, so ask them to justify it to those who are not sure.

$$\text{speed} = \frac{\text{circumference}}{T} = \frac{2\pi r}{T} = r \cdot \frac{(2\pi)}{T}$$

therefore $v = r\omega$.

The **actual** speed should be calculated by one method or the other and the units should be seen to be in m s^{-1}.

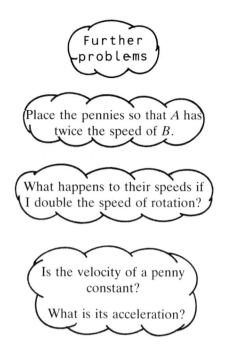
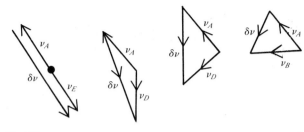

Fig 9.3

Place the pennies so that A has twice the speed of B.

As δt tends to zero, the textbook argument is that $\delta \mathbf{v}$ tends towards the radially inwards direction, and

$$\frac{\delta v}{\delta t} = \frac{r\omega \sin(\omega \delta t)}{\delta t} \simeq r\omega^2 \text{ since for small } \theta \sin \theta \simeq \theta$$

What happens to their speeds if I double the speed of rotation?

Is the velocity of a penny constant?

What is its acceleration?

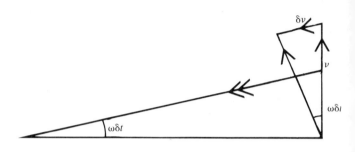

It is crucial to clarify the distinction between velocity and speed. The speed is constant but the velocity is changing. Clearly since acceleration is given by

$$\mathbf{a} = \frac{d\mathbf{\ddot{v}}}{dt}; \mathbf{a} \text{ is not zero!}$$

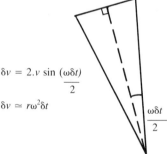

$$\delta v = 2.v \sin\left(\frac{\omega \delta t}{2}\right)$$

$$\delta v \simeq r\omega^2 \delta t$$

Problem

How is its velocity changing?

What is its acceleration?

Fig 9.4

Acceleration, therefore, acts radially inwards with magnitude $r\omega^2$ or $\dfrac{v^2}{r}$.

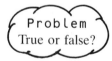
Problem
True or false?

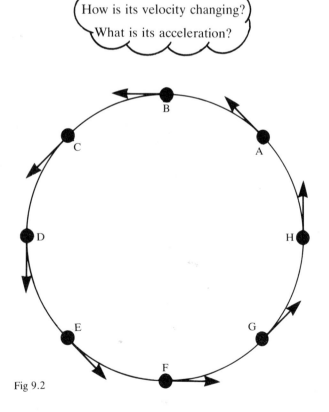

Fig 9.2

This section of the worksheet tests a student's understanding of the difference between the vector quantities displacement, velocity, acceleration and the scalar quantities, distance, speed, 'acceleration'.

Do note that, although acceleration is varying, its magnitude $r\omega^2$ remains constant. The last question introduces 'force' for the first time and discussion may lead into the dynamics of circular motion.

You can check that the students can subtract vectors by asking for the change in velocity $\mathbf{v}$ between position A and E; A and D; A and C; A and B.

Solution to the extension problem

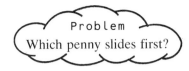

Problem
Which penny slides first?

Set up model   1

Assume that the penny is a particle of mass m. Let it be placed at a distance r from the axis of a rough, horizontal turntable which rotates with angular speed ω. Assume that friction between the penny and the turntable obeys the friction law, $F \leqslant \mu N$ where μ is the coefficient of static friction. The problem is now to find the critical angular speed at which the penny slides, i.e. for which $F = \mu N$.

Analyse the problem   2

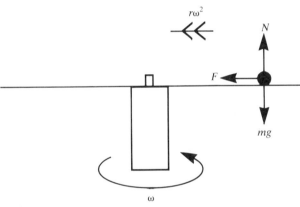

Fig 9.5

Apply Newton's second law in both the radial and vertical directions.

$$F = mr\omega^2$$
$$N - mg = 0$$

also

$$F = \mu N$$

when friction is limiting. Therefore

$$mr\omega^2 = \mu mg$$

and we have

$$\omega = \sqrt{\frac{\mu g}{r}}.$$

Interpret and validate [3]

1 Since $\omega \propto r^{-\frac{1}{2}}$, a penny which is further from the axis will have a smaller critical angular speed and will therefore slide first. This can be readily validated.

2 Calculations involving the measurement of ω, μ and r can be used to validate the theory. The coefficient of friction must be measured by tilting the turntable until the penny slides. (See chapter 7, 'Angle of friction'.)

Fig 9.6

3 Since ω is independent of mass, different coins at the same radius with the same coefficient of friction should slide at the same time. This is extremely difficult to validate because it is almost impossible to control μ and r.

Experiments often tend to show that ω is **not** mass-independent but that a larger mass has a smaller critical speed than a smaller mass in the same position. This cannot be explained using the present model. The model needs to be refined, in this instance by recognising that the linear friction law $F \leqslant \mu N$ is an approximate experimental law, which takes no account of the fact that μ decreases with increasing mass (see chapter 7, The Law of Friction).

😖 Misconceptions

The common misconceptions relating to circular motion are described in chapter 4, section 4.3. In particular, the plan above focusses upon the problems students have with the vector quality of velocity and acceleration.

Extensions

Investigate the path of a penny once it begins to slide

(a) relative to the turntable.
(b) relative to the ground.

Use talc or chalk-dust on the turntable and the sliding penny will leave a trace.

Conical pendulum (1)

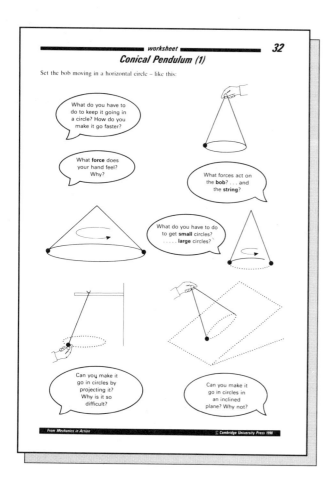

worksheet **32**
Conical Pendulum (1)

Set the bob moving in a horizontal circle – like this:

What do you have to do to keep it going in a circle? How do you make it go faster?

What **force** does your hand feel? Why?

What forces act on the **bob**? . . . and the **string**?

What do you have to do to get **small** circles? . . . . . **large** circles?

Can you make it go in circles by projecting it? Why is it so difficult?

Can you make it go in circles in an inclined plane? Why not?

From Mechanics in Action
© Cambridge University Press 1990

Aims

- To introduce or revise the concepts and theory of circular motion, with particular reference to the forces acting, i.e. the dynamics.
- To stimulate curiosity and thereby motivate the study of the conical pendulum.

Equipment

Provide fishing line, string, fishing weights and other masses to attach.

Plan

The worksheets and equipment should be provided for pairs of students. Encourage the students to discuss the questions raised and make a note of anything they find out or can explain. You will want to gather the groups together to extract all their discoveries and comments which might include the following.

- The mass slows down and eventually comes to rest unless you 'pump it' by rotating your hand.
- The faster you pump it, the higher the circles and the larger their radius.
- The faster the motion the greater the pull on the hand.

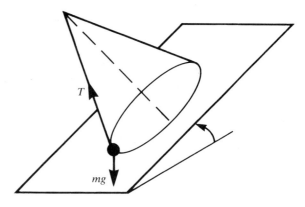

Fig 9.7

- For the same angular speed, **longer** strings make larger circles.
- There are two forces on the bob, its weight and the pull of the string.
- You need **just the right speed** of projection to get the bob going in a circle.
- It is not possible to get a circle in an inclined plane.
- There is a critical speed below which no horizontal circles are possible.

Solutions

See 'Conical pendulum (2)' for a solution to the problem 'How does the angle of the string depend on the angular speed?'

Extensions

1 The discussion should provide a useful starting point *either* for exposition of the essentials of the theory of circular motion, *or* for the identification of a number of interesting features to explain and problems to solve.

   Apart from the problems posed in 'Conical pendulum (2)' and 'Chairoplanes', here are two others.

2 Why does a conical pendulum rotate more quickly than a simple pendulum of the same length?
3 What happens to the mass, M, as you speed up the circular motion?

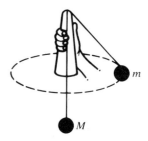

Fig 9.8

Banking

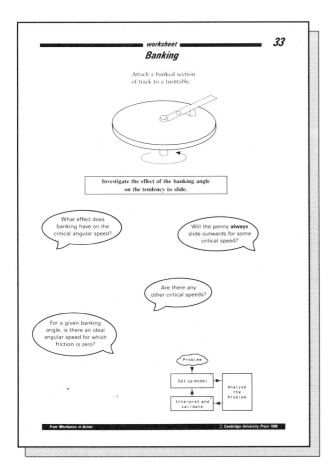

Equipment

A turntable with variable speed motor and banked
slideway, as provided by Unilab in the Leeds
Mechanics Kit. Stopwatch and adhesive tape.

Plan

Assuming a class demonstration, you will be able to
draw out the following features and problems.

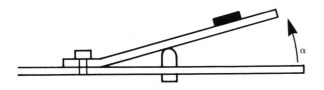

Fig 9.9

■ As in 'Pennies on a turntable', the outer pennies
slide first and quite high speeds are required to get
the inner pennies to slide.
■ The banking angle and friction both affect the critical speed, but what is the relationship?

■ At some angles, the penny will slide **down** the slide-
way if the angular speed is too low: there may be
two critical speeds. At which angles are there two,
one or no critical speeds?
■ At very high angles, even the **top speed** of the motor
is not sufficient to cause sliding. Is this because the
motor is too slow or is there **no** critical speed?

Solutions

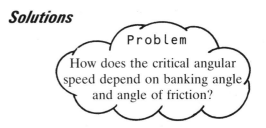

Problem
How does the critical angular
speed depend on banking angle
and angle of friction?

Set up model   1

Assume the penny is a particle of mass m, at a distance
r from the axis of the turntable which rotates at speed
ω. Friction is modelled in the usual way, $F \leqslant \mu N$,
where $\mu = \tan \lambda$.

Analyse the problem   2

Apply Newton's second law radially and vertically.

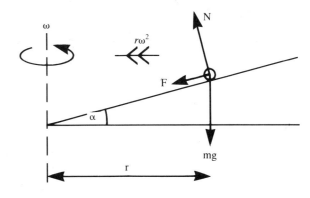

Fig 9.10

$$F \cos \alpha + N \sin \alpha = mr\omega^2$$

$$N \cos \alpha - F \sin \alpha = mg$$

and

$$-\mu N \leqslant F \leqslant \mu N.$$

Now, eliminating N and F gives

$$\frac{g}{r} \tan(\alpha - \lambda) \leqslant \omega^2 \leqslant \frac{g}{r} \tan(\alpha + \lambda)$$

or $\dfrac{g}{r}\left(\dfrac{\tan \alpha - \mu}{1 + \mu \tan \alpha}\right) \leqslant \omega^2 \leqslant \dfrac{g}{r}\left(\dfrac{\tan \alpha + \mu}{1 - \mu \tan \alpha}\right)$

1 The upper limit $\frac{g}{r}\tan(\alpha + \lambda)$ is only reached if

$(\alpha + \lambda) < \frac{\pi}{2}$ and clearly this critical outward sliding

speed **increases** with α, μ and decreases with r.

2 The lower limit is only reached if $\alpha - \lambda > 0$, i.e. if the banking angle is greater than the angle of friction. (Compare with chapter 7, 'Angle of friction'.)

3 Fig. 9.11 shows the values of α, λ for which there are 0, 1 or 2 critical speeds:

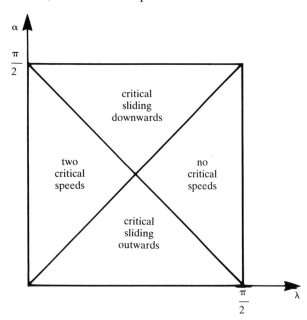

Fig 9.11

4 For fixed λ Fig. 9.12 shows the range of values of ω^2 for which the penny is at rest relative to the slideway.

Clearly the three parallel curves $\frac{g}{r}\tan(\alpha - \lambda)$, $\frac{g}{r}\tan\alpha$ and $\frac{g}{r}\tan(\alpha + \lambda)$ give the lower limit, the ideal speed for which friction is zero ($\lambda = 0$), and the upper limit, respectively.

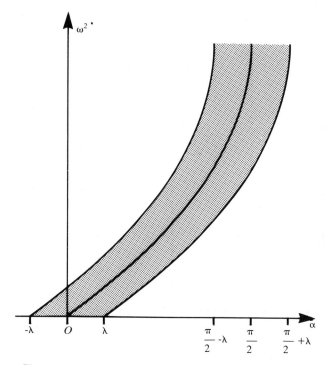

Fig 9.12

The rotor

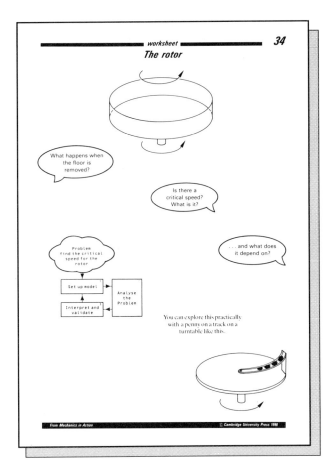

Equipment

A turntable with variable speed motor and rotor slide-way, as provided by Unilab in the Leeds Mechanics Kit. Stopwatches.

Plan

See the introduction to this chapter.

Solutions

Problem

What happens when the floor is removed?
Is there a critical speed?

Set up model   1

Assume the body is a particle of mass m, rotating with angular speed ω, in a horizontal circle of radius r. Let F and R be the friction and normal reaction exerted by the wall. Assume a coefficient of friction μ for contact between the body and the wall.

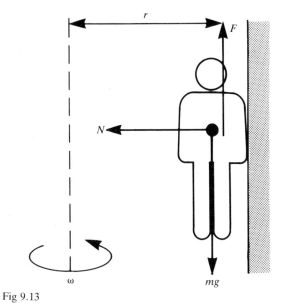

Fig 9.13

Analyse the problem   2

Applying Newton's second law in the vertical and radial directions

$$F - mg = 0$$

$$N = mr\omega^2.$$

The friction law $\dfrac{F}{N} \leq \mu$ gives $\dfrac{mg}{mr\omega^2} \leq \mu$

which gives

$$\omega^2 \geq \frac{g}{r\mu}.$$

1 With the floor removed the body will remain at rest relative to the rotor provided

$$\omega > \sqrt{\frac{g}{r\mu}}.$$

2 The critical speed is that at which friction is limiting, i.e. $F = \mu R$, and

$$\omega_C = \sqrt{\frac{g}{r\mu}}.$$

3 If the diameter is halved, the radius is also halved and the critical speed increased by a factor of $\sqrt{2}$.

4 The governing equations are the same as those for a penny on a turntable with F and R interchanged. Here **F** balances $m\mathbf{g}$ and **R** is the force towards the centre.

5 As $\mu \to 0$, $\omega_C \to \alpha$: the smoother the surface of the contact, the greater the required critical speed.

Extensions

What if I throw a ball to my friend on the opposite side?

 A misconception here is to think that A need only direct the throw along AB in order to reach B – as when the drum is at rest!

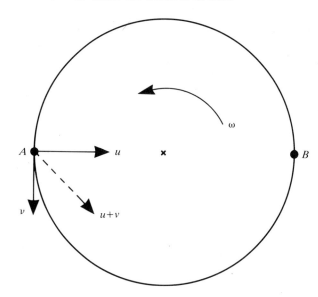

Fig 9.14

Conical pendulum (2)

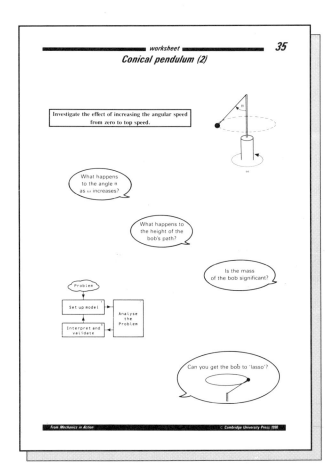

worksheet **35**
Conical pendulum (2)

Equipment

Variable speed motor with conical pendulum attachment as provided by Unilab in the Leeds Mechanics Kit. Alternatively, use fishing line and masses as in 'Conical pendulum (1)'.

Plan

See the introduction to this chapter.

Solutions

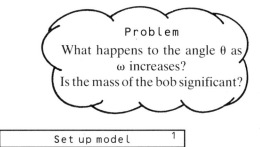

Problem
What happens to the angle θ as ω increases?
Is the mass of the bob significant?

| Set up model | 1 |

Assume the bob is a particle of mass m attached to a light and inextensible string of length l. Assume that the radius of the spindle is negligible so that the radius of the horizontal circle is $l \sin \theta$.
Forces acting on the bob are tension **T** and weight **mg**.

| Analyse the problem | 2 |

Apply Newton's second law in the vertical and radial directions

$$T \cos \theta - mg = 0$$
$$T \sin \theta = m(l \sin \theta)\omega^2$$

and therefore

$$T = ml\omega^2 \text{ or } \sin \theta = 0.$$

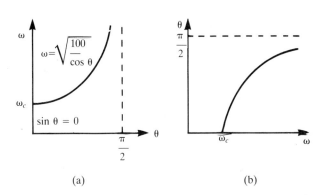

Fig 9.15

The solution can be expressed in the form

$$l \cos \theta = \frac{g}{\omega^2}$$

or

$$\sin \theta = 0$$

or

$$\cos \theta = \frac{g}{l\omega^2}$$

or

$$\sin \theta = 0 \qquad (9.2)$$

| Interpret and validate | 3 |

1 It follows from equation (9.2) that, as ω increases, cos θ decreases and therefore θ increases: the faster the angular speed the greater the angle and the more the pendulum swings out. As an illustration take $g = 10$ and $l = 0.1$ say, and enter

$$y = \sqrt{\frac{100}{\cos x}} \quad \text{or} \quad y = \cos^{-1}\left(\frac{100}{x^2}\right)$$

into a graphic calculator to produce two graphs.
   These equations describe ω as a function of θ and θ as a function of ω and are shown in Figs. 9.16(a) and (b) together with θ = 0, the solution corresponding to sin θ = 0 in equation (9.2).

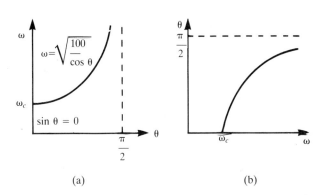

(a)   (b)

Fig 9.16

127

2 θ never reaches 90° however large ω becomes. As ω→∞, cos θ→0 and θ→90°. Clearly the lasso effect is impossible! An interesting problem now arises: 'How do cowboys get their ropes to lasso?'.

3 The graphs illustrate a critical angular speed ω_C below which the angle θ is zero. The bob only moves in circles when $\omega > \omega_C$ where $\omega_C = \sqrt{\dfrac{g}{l}}$. This

arises because cos θ ≤ 1 and equation (9.2) implies

$$\frac{g}{l\omega^2} \leq 1, \text{ i.e. } \omega \geq \sqrt{\frac{g}{l}}.$$

Validate experimentally using two bobs attached to two strings of length l_1 and l_2 ($l_2 > l_1$) and angular speed ω in the range

$$\frac{g}{l_2} < \omega^2 < \frac{g}{l_1}.$$

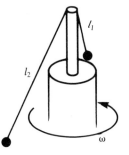

Fig 9.17

One bob, attached to the long string, moves in a circle whereas the other rests on the spindle with the string almost vertical and θ approximately zero. However this is **not** the θ = 0 solution since this only arises if the radius of the spindle, *r* is identically zero (see **8**). In practice the bob **will move in a small circle** when $\omega < \sqrt{\dfrac{g}{l}}$ provided it is not impeded by the spindle. This can be observed by turning the motor unit upside down (see Fig. 9.18).

The conical pendulum model is inappropriate and *r* must be taken into account as in the chairo-planes model (see 'Chairo-planes').

Fig 9.18

4 θ is independent of mass *m*. For a given angular speed the angle will be the same for bobs of different mass attached to strings of equal length.
Validate as shown in Fig. 9.19.

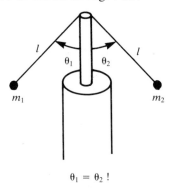

$\theta_1 = \theta_2$!

Fig 9.19

5 θ increases with length *l*. This follows from (9.2) or by drawing graphs of θ against ω for increasing *l* as in Fig. 9.20. Validate by setting up the apparatus with two strings of length l_1 and l_2 (see Fig. 9.21).

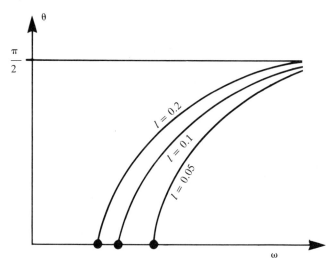

Fig 9.20

Observation of the motion in Fig. 9.21 reveals a startling fact. Although $\theta_2 > \theta_1$, $l_1 \cos \theta_1 = l_2 \cos \theta_2$, i.e. the two bobs move in circles in the **same** horizontal plane.

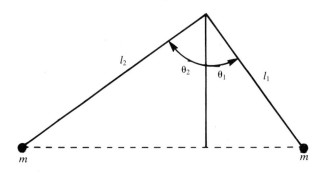

Fig 9.21

Equation (9.2) confirms that the depth of each bob below the apex is the same

$$l \cos \theta = \frac{g}{\omega^2}$$

so $l_1 \cos \theta_1 = l_2 \cos \theta_2 = \dfrac{g}{\omega^2}$.

6 A conical pendulum will always complete a circle more quickly than a **simple pendulum**, of the same length, completes an oscillation; this is because $\omega = \sqrt{\dfrac{g}{l}}$ for a simple pendulum.

7 $T = ml\omega^2$; the tension in the string increases with mass, length and angular speed. An examination of the breaking tension of the string will give a maximum safe mass, length or speed for the experiment.

8 A physical system which clearly demonstrates the $\sin \theta = 0$ solution when $\omega < \omega_C$ consists of a marble on the inside of a smooth bowl which rotates. As before

$$l \cos \theta = \frac{g}{\omega^2} \quad \text{or} \quad \sin \theta = 0.$$

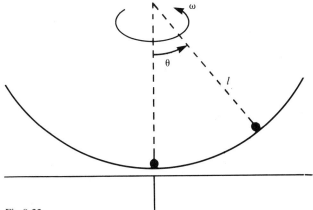

Fig 9.22

Extensions

New problems to investigate:

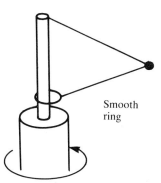

Smooth ring

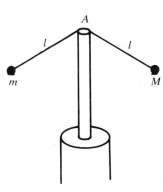

leave untied at A

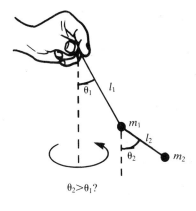

$\theta_2 > \theta_1$?

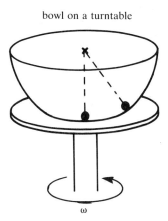

bowl on a turntable

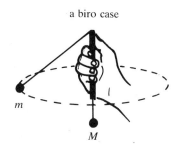

a biro case

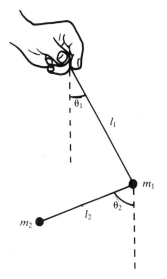

Fig 9.23

129

Chairoplanes

worksheet
Chairoplanes
36

Is the mass in the chair significant?

What happens to the angle θ as speed ω increases?

What about the length of the string?

Set up model

Analyse the Problem

Interpret and validate

Do inner and outer chairs swing out at the same angle?

From Mechanics in Action   © Cambridge University Press 1990

Equipment

Attach bob on string and tie on to the turntable driven by a variable speed motor such as that provided by Unilab in the Leeds Mechanics Kit.

Solutions

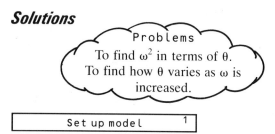

Problems
To find ω^2 in terms of θ.
To find how θ varies as ω is increased.

Set up model 1

As for the conical pendulum. Let the radius of the disc be r so that the chairs move in circles of radius $R = r + l \sin \theta$.

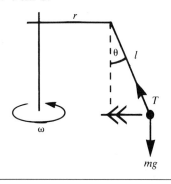

Analyse the problem 2

Newton's second law applied vertically and radially

$$T \cos \theta - mg = 0$$

$$T \sin \theta = m(r + l \sin \theta)\, \omega^2.$$

Solution

$$\omega^2 = \frac{g}{(l \cos \theta + r \cot \theta)}. \qquad (9.3)$$

Interpret and validate 3

1 Taking $g = 10$, $l = 0.1$ and $\dfrac{r}{l} = \mu$

then

$$\omega^2 = \frac{100}{(\cos \theta + \mu \cot \theta)}$$

which can be sketched on a graphic calculator for various values of μ, see Fig. 9.24.

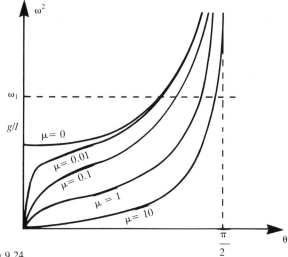

Fig 9.24

As with the conical pendulum it follows that θ is independent of mass and increases with ω and l. In fact as $\omega \to \infty$, $\cos \theta \to 0$, i.e. $\theta \to \dfrac{\pi}{2}$.

2 For fixed l and fixed ω ($\omega = \omega_1$ say) then the graphs in Fig. 9.24 show that θ increases with μ, i.e. with r. Consequently inner chairs swing out at a **smaller** angle than outer chairs having the same l.

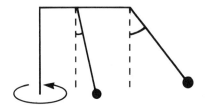

3 As $r \to 0$ equation (9.3) suggests that the conical pendulum solution, $\omega^2 = \dfrac{g}{l\cos\theta}$ is recovered. In fact closer examination of the graphs of

$$\omega^2 = \frac{g}{l(\cos\theta + \mu\cot\theta)}$$

shows that as $\mu \to 0$, the graph does **not** approach $\omega^2 = \dfrac{g}{l\cos\theta}$ near the origin. Indeed as $\omega \to 0$, $\theta \to 0$ for any non-zero μ, no matter how small.

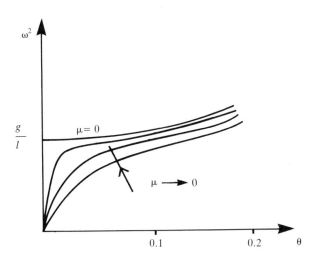

Fig 9.25

This shows that there is no critical speed, $\sqrt{\dfrac{g}{l}}$, for the chairoplanes and that circular motion arises for any ω, no matter how small the radius r compared to length l.

4 When $r \gg l$, $\mu \to \infty$ and the graph looks increasingly like that of a 'tan function'. Indeed from equation (9.2)

$$\omega^2 = \frac{g}{l\cos\theta + r\cos\theta} = \frac{g\tan\theta}{r(1 + \dfrac{l}{r}\sin\theta)}$$

$$\to \frac{g\tan\theta}{r} \text{ as } r \to \infty.$$

This model may be recognised as that which describes the motion of a particle on a **smooth**, banked track in which reaction N replaces tension. (See 'Banking' in this chapter.)

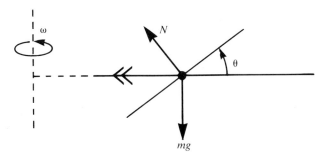

5 The depth below the apex, $d = l\cos\theta$ in the conical pendulum, is here replaced by a depth

$$D = l\cos\theta + r\cot\theta.$$

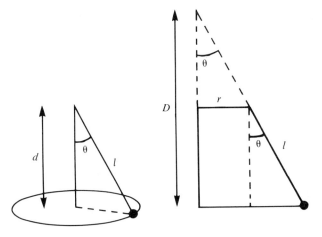

Fig 9.26

Since $D = \dfrac{g}{\omega^2}$ all chairs have the same depth D no matter what their r and l values. But, unlike the conical pendulum, they do not necessarily rotate in the same horizontal plane. Figure 9.27 shows two chairs with the same depth D. When the chairs have the same l and different radii (r_1, r_2), geometric considerations confirm that $\theta_2 > \theta_1$.

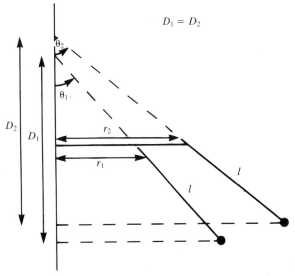

Fig 9.27

Looping the loop (2)

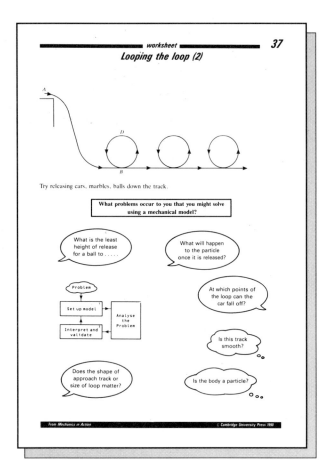

Equipment

'Streak' car racing track, cars, balls and marbles.

Plan

It is assumed that students have met the theory of circular motion and the conservation of energy. They may have seen the apparatus in the context of 'Looping the loop (1)' in chapter 8.

There are a number of points which you can raise which will help your students to tackle the problems.

1 The **difference** between the performance of a car and a marble will draw attention to the role of energy losses and friction in the problem. So also will the difference between release heights for one loop and two loops.
2 Students often think the critical condition for complete loops is $h = 2a$, i.e. the least speed at the top of the loop is zero. This can be disproved in practice; when the ball just completes a loop, it can be observed to have a significant speed at the top of the loop.
3 It is often possible to see the track shake or move as the car goes round a loop: this can be explained in terms of the reaction between car and track and the change in momentum involved. It is also possi-

ble to press down on the outside of the track and 'feel the motion' of the car as it passes. Discussion can thereby be focused on the crucial issue of the reaction force between car and track.

Solutions

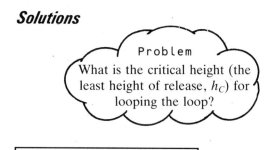

Problem

What is the critical height (the least height of release, h_C) for looping the loop?

Set up model	1

Consider a simple model in which the car/ball is assumed to be a particle of mass m and that friction is negligible so that energy is conserved. Assume the loop is a circle of radius a and the height of release is h above the lowest point, B.

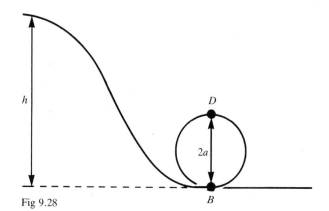

Fig 9.28

Analyse the problem	2

If U is the speed of the particle at B then conservation of energy gives
$$\tfrac{1}{2}mU^2 = mgh \implies U^2 = 2gh. \tag{9.4}$$
Students are likely to argue that if the particle reaches the top of the loop, D, with speed V then by conservation of energy
$$\tfrac{1}{2}mU^2 = \tfrac{1}{2}mV^2 + mg2a.$$
Forces acting are mg and reaction N vertically downwards and so Newton's second law gives
$$mg + N = \frac{mV^2}{a}$$
$$\implies \frac{N}{mg} = \frac{U^2}{ag} - 5 = \frac{2h}{a} - 5$$
from (9.4).

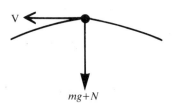

Particle is at the top of the curve

The particle will therefore reach D if $N \geqslant 0$,

i.e. $h \geqslant \frac{5}{2}a$.

This simple argument is correct but insufficiently rigorous! We need to prove that $N > 0$ all round the track (from B to D and on to B again) by first determining N at a general position C where the speed is v;

$$N - mg\cos\theta = \frac{mv^2}{a}$$

therefore

$$N = 2g\left(\frac{h}{a} - 1\right) + 3g\cos\theta.$$

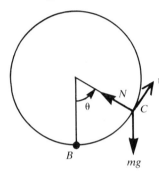

Fig 9.29

Now impose the condition, $N \geqslant 0$ for all θ, in order to loop the loop! In particular the **minimum value** of N occurs when $\theta = \pi$

$$N_{\min} = 2g\left(\frac{h}{a} - 1\right) - 3g \geqslant 0,$$

so

$$h \geqslant \frac{5a}{2}, \quad h_C = \frac{5a}{2}.$$

Interpret and validate 3

1 At which points of the loop can the car fall off? Since

$$R = \frac{mv^2}{a} + mg\cos\theta \quad \text{then} \quad R \geqslant 0 \quad \text{for} \quad 0 \leqslant \theta \leqslant \tfrac{1}{2}\pi.$$

Falling off can only arise when $\theta > \dfrac{\pi}{2}$ and in particular when $\cos\theta = \dfrac{2}{3}\left(1 - \dfrac{h}{a}\right)$.

The particle will fall off the track if the height of release h is given by

$$a < h < \frac{5a}{2}.$$

2 When students try to validate the result, $h_C = \dfrac{5a}{2}$ they invariably find large errors (of the order of 100%) using cars and smaller errors (of the order of 20% or more) using marbles. This suggests that the simple model which assumes the car/marble to be a particle and the track to be smooth is not very good. A more refined model is therefore required which takes account of friction and the finite size of the marble.

Refine the model

Assume the marble is a sphere of radius b and mass m. Assume that the marble **rolls** along and around the rough track. Consequently the point of contact between the marble and the track will be instantaneously at rest (see chapter 1, section 1.2(f)) and friction does **no work**. Therefore energy is conserved. Kinetic energy now includes rotational and translational components (see chapter 1, section 1.2 (f))

$$KE = \tfrac{1}{2}mv^2 + \tfrac{1}{2}I\omega^2$$
$$KE = \tfrac{7}{10}mv^2$$

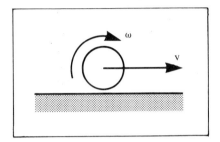

and an improved value for critical height is

$$h_C = 2.7a \quad \text{as} \quad \frac{b}{a} \to 0.$$

This prediction is now closer to experiment yet some error still remains – at least 10–15%! This can be traced to energy losses due to jerks (where two pieces of track are joined) and **sliding**. The ball will fail to roll and will therefore slide if the track becomes too steep. Clearly, even if the marble rolls along the approach track, sliding is bound to occur when negotiating the loop. The extent of this energy loss can be seen by setting up two loops and having the marble just complete the first loop, from height h_1; then both loops from height h_2.

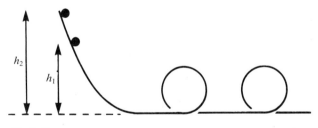

Fig 9.30

😦 Misconceptions

The most frequent misconception is to assume that the particle will just loop the loop if it arrives at D with zero speed, $h_C = 2a$.

Extensions

1 Once the particle leaves the track it travels as a projectile; what is the maximum height attained and where does it land?
2 On the Corkscrew, the train is actually **all** going at the same speed. It is possible to model this as a series of connected particles. Find out whether this is more or less likely to fall off the track and whether the front of the train is more or less dangerous than the rear.

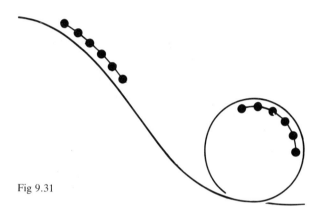

Fig 9.31

3 For a given coefficient of friction, μ, find the least angle of slope of the approach track for which a ball will slide. (For a particle and a rolling sphere, the angles are $\tan^{-1}\mu$ and $\tan^{-1}\frac{7}{2}\mu$ respectively.)

Wind-up

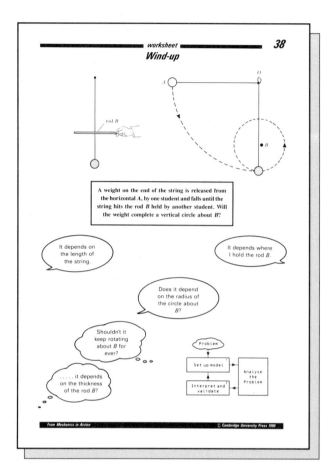

Equipment

String, bob and pencil.

Plan

See the introduction to this chapter.

Solution

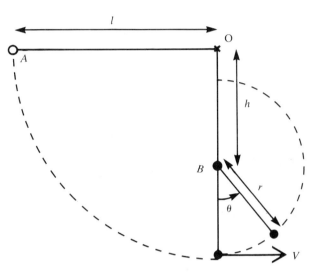

Problem
Find the critical depth of B below O for the particle to complete a circle.

Set up model	1

Fig 9.32

A particle of mass m is attached to an inextensible string of length l. Assume that it falls from rest at A and then rotates in a circle about O. When it is vertically below O its speed is V and the string now makes contact with the rod B; assume $OB = h$ and the diameter of rod is negligible.

Analyse the problem	2

Conservation of energy gives $V^2 = 2gl$ and the usual analysis for motion in a vertical circle provides a condition for the particle to complete a circle about B;

$$V^2 \geqslant 5gr \text{ where } r = l - h.$$

Therefore

$$2l \geqslant 5(l - h)$$

and the solution is

$$h \geqslant \tfrac{3}{5}l$$

Interpret and validate	3

Measurements should validate the result. If the depth of B below O is greater than $\frac{3}{5}l$, then vertical circles will occur. If $OB < \frac{3}{5}l$, the string will go slack in some position and a full circle will not be completed.

Extensions

1 Note that both air resistance and the thickness of the rod at B are ignored. Air resistance tends to reduce the number of completed circles and the thickness of the rod increases this number. A refined model could incorporate the radius of the rod.

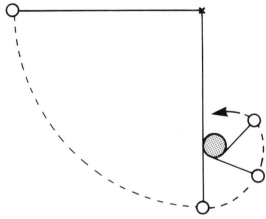

Fig 9.33

The problem then is 'Will the string completely wind up around the rod?'

2 A bar game allows you to release the ball from any angle θ. If the aim of the game is to strike the rod B with the 'particle', what value of θ should you choose?

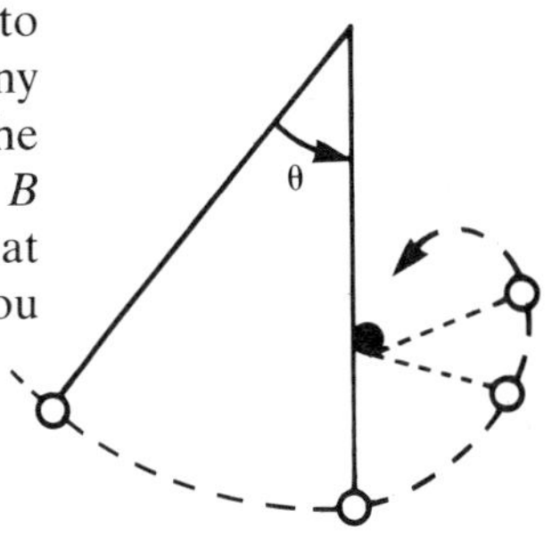

Fig 9.34

3 What is the least speed with which the 'water in the bucket' trick can be performed?

Is it best to have long or short arms?

Fig 9.35

A stop–go phenomenon

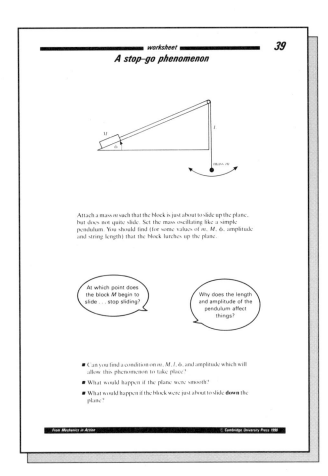

worksheet ■ **39**
A stop–go phenomenon

Attach a mass *m* such that the block is just about to slide up the plane, but does not quite slide. Set the mass oscillating like a simple pendulum. You should find (for some values of *m, M, φ*, amplitude and string length) that the block lurches up the plane.

At which point does the block *M* begin to slide . . . stop sliding?

Why does the length and amplitude of the pendulum affect things?

■ Can you find a condition on *m, M, l, φ*, and amplitude which will allow this phenomenon to take place?
■ What would happen if the plane were smooth?
■ What would happen if the block were just about to slide **down** the plane?

From Mechanics in Action   © Cambridge University Press 1990

Equipment

An inclined plane, blocks, strings and masses.

Plan

Arrange the system so that the mass *m* is **nearly** large enough to pull the block of mass *M* up the inclined plane. Then set the mass *m* oscillating and observe the block move up the plane, stop, and move on again, stop again, etc. You may need to adjust the angle of incline slightly. Try various amplitudes and lengths of pendulum.

Discuss with the whole class:

■ how does the tension in the string vary when the pendulum swings? What is its maximum and minimum?
■ what tension is required to drag the mass up the plane? (What tension is required to stop the mass sliding down the plane?)

Qualitative descriptions of the forces involved and the motion arising should now provide quantitative descriptions, that is, introduction of variables and application of principles. Set the problem for groups or individuals to tackle. Extensions will provide very challenging modelling investigations.

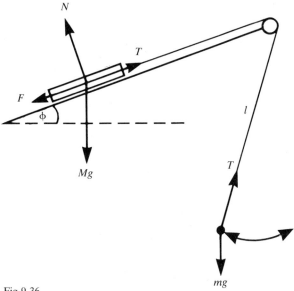

Fig 9.36

Solution

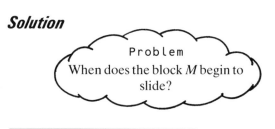

Problem
When does the block *M* begin to slide?

| Set up model | 1 |

Assume the friction law $F \leq \mu N$ for the block. Assume the pulley is smooth and the string is light and inextensible.

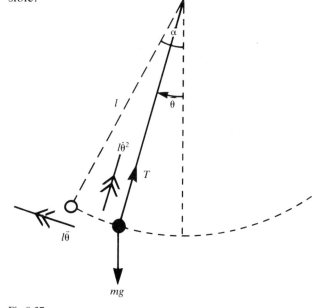

Fig 9.37

| Analyse the problem | 2 |

Newton's second law for the pendulum gives

$$T - mg\cos\theta = ml\dot{\theta}^2,$$
$$-mg\sin\theta = ml\ddot{\theta}.$$

Alternatively energy conservation gives

$$\tfrac{1}{2}m\,(l\dot{\theta})^2 = mg\,l\,(\cos\theta - \cos\alpha)$$

where α is the amplitude.

Therefore,

$$T = mg\,(3\cos\theta - 2\cos\alpha).$$

Analysing the forces acting on the block as in Least Force Problems (1) (See chapter 7), the block will slide if

$$T = \mu Mg\cos\alpha + Mg\sin\alpha$$
$$= \frac{Mg\,\sin\,(\alpha + \lambda)}{\cos\lambda}.$$

Therefore sliding starts when

$$mg\,(3\cos\theta - 2\cos\alpha) = Mg\,\frac{\sin\,(\alpha + \lambda)}{\cos\lambda}.$$

If the range of tensions given by

$$T = mg\,(3\cos\theta - 2\cos\alpha)$$

is greater than the value of

$$Mg\,\frac{\sin\,(\alpha + \lambda)}{\cos\lambda}$$

then the block will start and continue to slide up the plane. In particular if

$$T = mg\cos\alpha > Mg\,\frac{\sin\,(\alpha + \lambda)}{\cos\lambda}$$

then the block will never be in equilibrium.

Note   Replacing λ by $-\lambda$ gives the condition for preventing the block sliding **down** the plane!

$$T > Mg\,\frac{\sin\,(\alpha - \lambda)}{\cos\lambda}.$$

Interpret and validate [3]

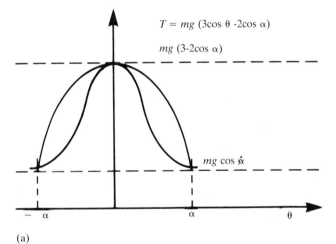

(a)

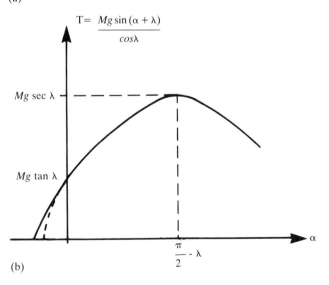

(b)

Fig 9.38

Cake tin

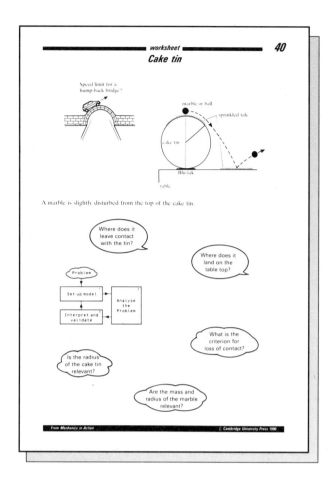

Equipment

A cake tin, marbles or rubber balls, Blu-tak or plasticine and talcum powder.

Plan

See the introduction to this chapter.

Solution

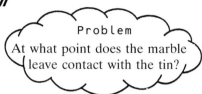

Problem
At what point does the marble leave contact with the tin?

| Set up model | 1 |

Assume the marble is a **particle**, initially at rest at A, and sliding on a **smooth** cylinder of radius a. The problem is to calculate the angle ϕ where the reaction force $N = 0$.

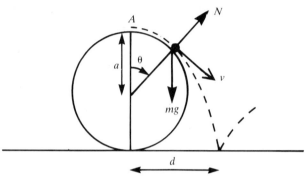

Fig 9.39

| Analyse the problem | 2 |

Newton's second law in the radial direction gives

$$mg \cos \theta - N = \frac{mv^2}{a}$$

Energy conservation gives

$$\tfrac{1}{2}mv^2 = mga\,(1 - \cos \theta).$$

Eliminating v gives

$$N = (3 \cos \theta - 2)\,mg.$$

Since $N = 0$ when $\theta = \phi$, then $\cos \phi = \tfrac{2}{3}$.

1 The angle at which the marble leaves contact is 48°,
   or $\cos^{-1}\left(\frac{2}{3}\right)$; it is probably easiest to check the depth
   below A. This can be validated in the practical. A
   refined theory incorporating the **rolling** energy of
   the marble

$$\text{KE} = \tfrac{7}{10}mv^2$$

gives $\phi = \cos^{-1}\left(\tfrac{10}{17}\right) = 54°$

2 The angle is independent of the radius of the tin
   and the mass of the marble. These interpretations
   can be investigated practically.

Extensions

1 Projectile theory can be used to calculate the distance d of the landing point from the tin. A particle
   model gives $d = 1.46a$. A rigid body model for a
   sphere of small radius gives $d = 1.78a$. (The analysis
   can be extended to a rolling sphere of significant
   radius.)
2 What is the greatest safe speed for a car to go over
   a hump-back bridge?

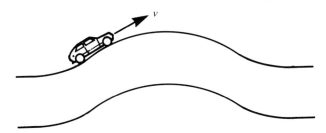

Fig 9.40

chapter
10
Springs, oscillations and simple harmonic motion

The practicals in this chapter can provide material for motivating topics in pure mathematics, for introducing new theory in mechanics or starting points for extended modelling investigations.

Timing oscillations can be used to collect data to model with functions and graphs. It also can be used to validate the theory of simple harmonic motion, and to provide a starting point for a modelling investigation of various oscillating systems. '**The simple pendulum**' later in the chapter can be used in a similar fashion.

Modelling a mass spring oscillator is an ideal way to introduce the theory of simple harmonic motion as a model of an oscillating system. All the key results of simple harmonic motion should arise in the investigation in a natural and meaningful way. The practical can also be used with pure mathematics students as an introduction to sine and cosine functions and their properties.

Hooke's law and Stiffness and elasticity are practicals which introduce Hooke's law as a **model** for tension in elastic strings and springs.

Amplitude decay and Resonance provide practicals which motivate the students to refine and extend the simple harmonic motion model for oscillations. These involve an investigation of new functions and differential equations and provide ample scope for motivating students' personal research and study. They can also be used by pure mathematics students without recourse to Newtonian mechanics simply as sources of data to model with functions and graphs.

The simple pendulum provides a straightforward example of a practical validation of the theory of simple harmonic motion. It illustrates the importance of bearing in mind the assumptions made in setting up the model. Pure mathematics students may collect data and model with functions and graphs. Statistics students may examine the hypotheses that time period is independent of mass and amplitude.

The compound pendulum is a practical and a modelling investigation which involves rigid body dynamics, including moments of inertia. As such, it may provide A-level mechanics students with the motivation to undertake some personal research.

Timing oscillations

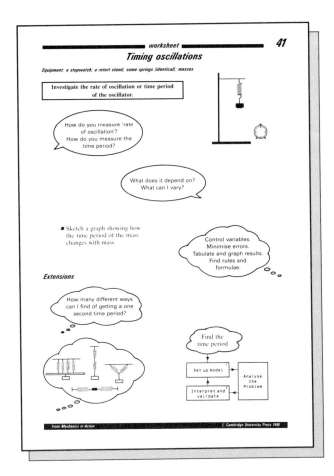

Aims

- To provide a modelling investigation using only pure mathematics – functions and graphs.
- To motivate the study of simple harmonic motion.
- To **validate** the theory of simple harmonic motion.
- To provide a starting point for an investigation which applies the theory of simple harmonic motion to various systems.

Equipment

Each group requires a set of springs and masses as provided by Unilab in the Leeds Mechanics Kit and in the Practical Mathematics Kit. It is helpful if some of the springs are identical, and if the masses provide a wide range and time periods. The choice of masses depends on the springs being used.

With a stiff spring, about eight 100 g masses, a 50 g mass and a holder will ensure a wide range of time periods. For a very elastic spring, a 100 g, 50 g and several 10 g masses might be appropriate. The important thing is that there is a wide range of time periods of oscillation available, for example from $T = 0.25$ s to $T = 2$ s. Each group may require a stopwatch, though ordinary watches are adequate, and many students do have wrist stopwatches.

A supply of graphic calculators or function graph plotters is desirable.

Plan A

A 70 minute lesson for lower sixth pure mathematics A-level students

Set the mass spring oscillator in motion.

'Estimate its time period.'
'Estimate its frequency, the number of oscillations per second.'

Now hand out some stopwatches and ask the students to time the oscillations. Time 10 oscillations, say. Notice the discrepancies between different answers.

'How accurate are our readings?'

Pose a problem:

'If I double the mass, what would happen to the time period? Guess.'

Test your guess, then set the investigation:

'Find a general rule for the variation of the time period or frequency with mass. Investigate other factors which may affect the time period or frequency.'

Organise the group into threes to tackle the investigation.

You should encourage the students to check the **consistency** and **accuracy** of measurements taken. If more than one student times the oscillations, or if measurements are repeated, what **variation** in measurements is obtained?

It is expected that most groups will collect data and draw a graph. Does it pass through the origin? Should it pass through the origin? They should be encouraged to try to find a formula. This might be done by hand by plotting the mass against the **square** of the frequency (or time period), for example. However, with the aid of a graphic calculator or function graph plotter, students will vary parameters until they obtain a graph of best fit. Alternative formulae might well arise; a straight line graph may provide a reasonable fit to the data although Newtonian theory (see 'Solution' below) gives a square root relationship for T against m.

Whatever formula or model is obtained, students should be encouraged to make predictions and test their validity in practice. **Interpolation** within the range of values for which data was collected should give valid predictions, though they should be asked to provide an estimate of the acceptable error of their prediction based on their estimate of accuracy and consistency in the data collected. For instance, if their data was accurate and consistent to within 10%, they may make predictions within 10% with some confidence.

Students should also be encouraged to **extrapolate**. It is unlikely that their formula will allow them to extrapolate far, since:

(a) the model may become increasingly inaccurate outside the range of data collected.
(b) the behaviour of the springs will change under extreme loads.

They should therefore be encouraged to find a range of values within which their models are valid.

The investigations of different groups should be reported to the class (probably on a subsequent occasion) and comparisons made.

Some groups may have varied the mass, some may have investigated the number of springs used in series or parallel and some may have investigated different amplitudes of oscillation. Comparisons of their reports may yield the following:

1 Different formulae applied to similar-looking data or even the same data.

The straight line model is a good local approximation of the non-linear model. In Fig. 10.1 extrapolation of the linear model will yield invalid predictions, though both models are adequate for interpolation.

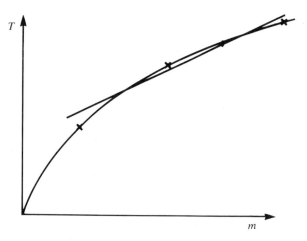

Fig. 10.1

2 Doubling the mass has the same effect as doubling the length of spring, by attaching two springs in series.

Plan B

30 minute practical validation of simple harmonic motion theory with an A-level applied mathematics class

The problem can be posed as in plan A; you need not assume that students will find this practical investigation trivial even when they have met the theory of simple harmonic motion previously. In plan B you will encourage the students to apply a modelling approach using Newton's laws and the theory of simple harmonic motion (see 'Solutions').

Note that validation of the theory may involve measuring the stiffness of the spring or spring system. This may involve investigating how this depends on the number of springs in series or in parallel. (See the notes in this chapter on 'Hooke's law'.)

An extended investigation may follow in which groups are asked to set up a variety of their own spring systems and to model them mathematically. For example, see Figs. 10.2 and 10.3.

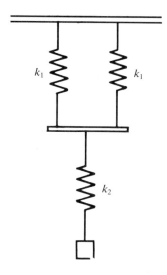

Fig. 10.2 How does the time period depend on k_1 and k_2?

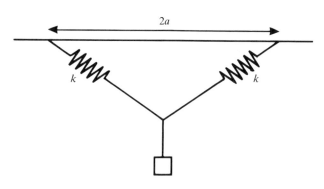

Fig. 10.3 How does the time period depend on a and k?

In cases where the Newtonian modelling proves difficult, some positive achievement can be gained from the empirical modelling with functions and graphs of the kind described in plan A.

Solutions

Here the Newtonian model is given for a simple mass spring oscillator.

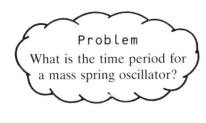

Problem
What is the time period for a mass spring oscillator?

Set up model	1

The time period, T, depends on the mass attached, m, and the spring stiffness, k. Assume the spring is 'Hookean' (see 'Hooke's law') and the motion takes place in a vertical line.

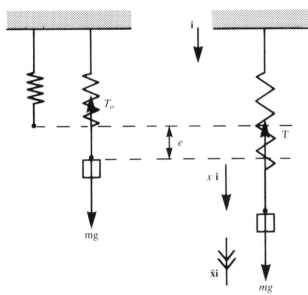

Fig. 10.4

Analyse the problem	2

When the mass hangs in equilibrium, let the extension be e and tension, T_0. Then Newton's second law gives

$$T_0 - mg = 0$$

and Hooke's law for the spring $T_0 = ke$.

Consequently $mg = ke$.

When the mass has displacement $x\mathbf{i}$ and acceleration $\ddot{x}\mathbf{i}$, let the tension be T.
Newton's second law gives

$$mg - T = m\ddot{x}$$

Hooke's law for the spring is

$$T = k\,(x + e).$$

Eliminating T we obtain the simple harmonic motion equation

$$m\ddot{x} = -kx$$

or $$m\ddot{x} + kx = 0.$$

Substituting $x = A \cos \omega t$ (or $A \sin \omega t$) into this equation gives

$$\omega^2 = \frac{k}{m}.$$

Therefore the solution is a cosine (or sine) wave with frequency $\omega = \sqrt{\dfrac{k}{m}}$ or time period $T = 2\pi \sqrt{\dfrac{m}{k}}$.

Interpret and validate	3

1. Actual measurements of m and k can be taken and predicted values of T tested.

2. The proportionality $T \propto \sqrt{m}$ can easily be validated, for example by quadrupling the mass and testing for a doubled time period.

3. The result $T \propto \sqrt{\dfrac{1}{k}}$ can be similarly tested, by doubling or halving the spring stiffness. This can be done by attaching two springs in parallel or series respectively. (See 'Hooke's law' notes.)

4. Clearly $x = A \cos \omega t$ is only approximately valid for a limited length of time. Eventually the amplitude A noticeably decays, due to energy lost to the air, friction at the point of support, or to heat in the spring itself. A refined model is given in the notes for 'Amplitude decay'.

5. The time period is independent of the amplitude! This result can also be validated.

6. Substituting $mg = ke$ into the formula gives $T = 2\pi \sqrt{\dfrac{e}{g}}$. It is clearly only necessary to know the extension of the spring when it is in equilibrium.

Extensions

1 Extensions to other spring systems can be suggested by the students. They should be aware that there are analytical difficulties involved in some cases! Here are some suggestions which are reasonable.

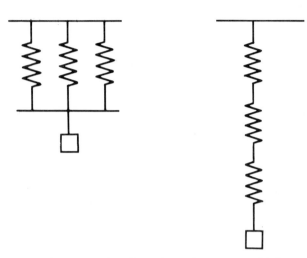

Fig. 10.5 For n springs in series and parallel $T \propto 1/\sqrt{n}$ and $T \propto \sqrt{n}$

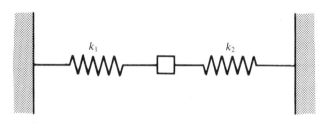

Fig. 10.6 Springs on a smooth horizontal plane. Time period T.
$$T \propto \sqrt{k_1 + k_2}$$

2 The analysis depends on Hooke's law being valid; however, if there is significant pretensioning in the spring, $T = T_0 + kx$ may be a better model. Students should investigate the effect of refining the model. In fact, $T = 2\pi \sqrt{m/k}$ is still the result for $m > T_0/g$, though this no longer gives $T = 2\pi\sqrt{e/g}$.

Other models for tension might also be looked at

$$T = T_0 + k_1x + k_2x^2 + k_3x^3 + \ldots$$

Modelling a mass spring oscillator

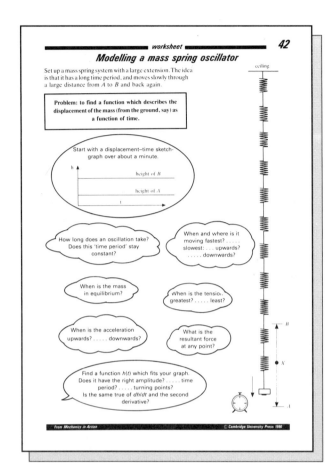

Aims

- To provide a periodic motion to investigate and describe using a sketch graph and an appropriate function.
- To introduce $x = A \cos \omega t$ as a function which models the oscillator's motion, and so to introduce the theory of simple harmonic motion.

Equipment

Provide enough springs and masses to set up a relatively slow oscillation of period about 2 seconds if possible. You may need a point of attachment to the ceiling. You also need a stopwatch and a metre ruler.

Plan

Set the mass oscillating and pose the problem given on the worksheet. The questions provided on the worksheet can be tackled together as a class. The following points should be brought out:
- the motion is 'periodic'.
- the velocity is given by the slope of the graph.
- the velocity is **zero** at A and B.

- the equilibrium position, X, should be half way between A and B. This can be checked by stopping the oscillation and measuring with a ruler.
- the dynamic equilibrium point, X, is the same as the static equilibrium point.
- the acceleration is zero at X; this implies the velocity is a maximum or minimum!
- acceleration is directed downwards when the mass is above X and upwards when the mass is below X.
- time can be measured from any instant, and displacement may be referenced from any point, in particular from the equilibrium point X.

These points might be noted on the board. The class should now be ready to try to sketch a graph; get each student individually to sketch an appropriate graph, and ask volunteers to put one on the blackboard for discussion.

Compare graphs. Obviously, sine and cosine are possible functions, but the following should also be noted.

(a) If displacement is measured **from the ground** then the function is

$$x = (\text{height of } X) + \frac{(AB)}{2} \cdot \text{cosine (?)}.$$

(b) If the mass is projected from X then sine is appropriate.
(c) If $2a = AB$ then the simple function

$$x(t) = a \cos t$$

has the correct amplitude but an incorrect period of $2\pi \sim 6.3$ seconds.

At this stage an important step in the investigation is to introduce a constant ω and try to model the displacement by the function

$$x(t) = a \cos \omega t.$$

Suppose we have $T = 2$ seconds, for instance, then:

$\omega = 1$; $x(t) = a \cos t$, period is too large, $T \sim 6.3$ s

$\omega = 2$; $x(t) = a \cos 2t$, period is too large, $T \sim 3.15$ s

$\omega = 3$; $x(t) = a \cos 3t$, period is too large, $T \sim 2.1$ s

$\omega = 4$; $x(t) = a \cos 4t$, period is too small, $T \sim 1.58$ s

$\omega = 3.15$; $x(t) = a \cos 3.15t$, period is $T = 2$ s!

Even more precise calculation of ω gives $\omega = \pi$ when $T = 2$.

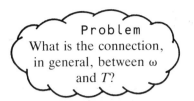

Problem
What is the connection, in general, between ω and T?

If $T =$ time for 1 oscillation, during this time of T seconds, ωt increases by 2π so that $x(t) = a \cos \omega t$ starts and finishes with the **same value**. Therefore $\omega T = 2\pi$,

$$\omega = \frac{2\pi}{T} \quad \text{or} \quad T = \frac{2\pi}{\omega}.$$

The class discussion and investigation should conclude with the results:

- $x = a \cos \omega t$;
- $\omega = \dfrac{2\pi}{T}$, where T is the time period in seconds;
- the graph which describes the motion is periodic.

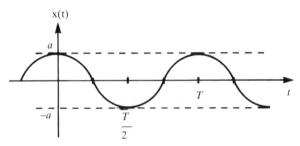

Fig 10.7

The formulae and graphs for $\dfrac{dx}{dt}$ and $\dfrac{d^2x}{dt^2}$ should now follow. The observation that $\dfrac{d^2x}{dt^2} = -\omega^2 x$ should now arise as a consequence. This provides an appropriate starting point for the study of simple harmonic motion.

Hooke's law

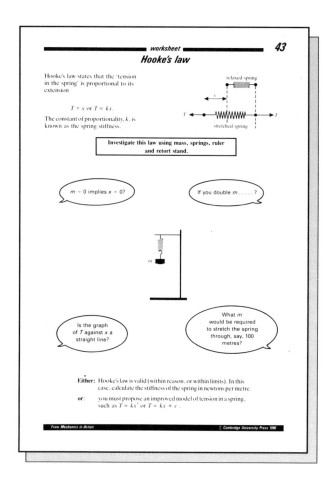

Aims

- To introduce Hooke's law as a model for tension in a spring, and the tension–extension gradient as a measure of the stiffness of a spring.
- To provide an appreciation of the approximate nature of the model and of its limitations.

Equipment

Springs of different stiffness and masses (10 g, 50 g and 100 g) are provided by Unilab in the Leeds Mechanics Kit and in the Practical Mathematics Kit.

Plan

Give each group a worksheet and a **different** spring. Try to arrange for the different groups to obtain springs of very different stiffnesses. Points to bring out when the results from different groups are collected are:

- springs are often pretensioned. In such cases Hooke's law will need to be refined to formulae such as $T(x) = T_0 + kx$.
- there is a **range** of extensions over which Hooke's law is approximately valid.

(See chapter 2 for more details.)

Stiffness and elasticity

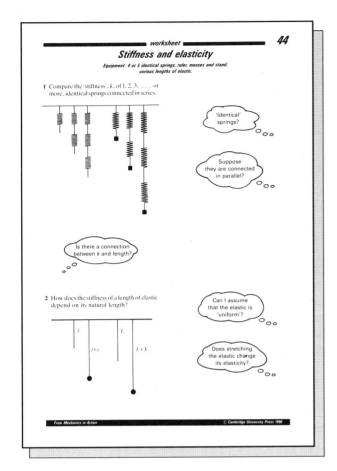

Aims

- To introduce elasticity as a property of the material independent of length.
- To extend the formula for Hooke's law to $T = \dfrac{\lambda x}{l}$, where l is the length of the string or spring.

Equipment

Note that some kinds of elastic have a plastic coating: when the elastic is **first** stretched, this coating breaks up and this affects the elasticity of the string. Subsequently the elasticity should hardly change unless it is stretched beyond its elastic limit.

Plan

Some groups can be given springs and others lengths of elastic. The aim is to investigate how stiffness varies with the **length** of the spring or string.

Solutions

If the elasticity of the springs and string does not vary too much, the stiffness should be found to vary inversely with the length of string or spring,

$$k \propto \frac{1}{l} \quad \text{or} \quad k = \frac{\lambda}{l}$$

Hooke's law, $T = kx$, becomes $T = \dfrac{\lambda x}{l}$

where λ is called the elasticity of the string or spring.

Amplitude decay

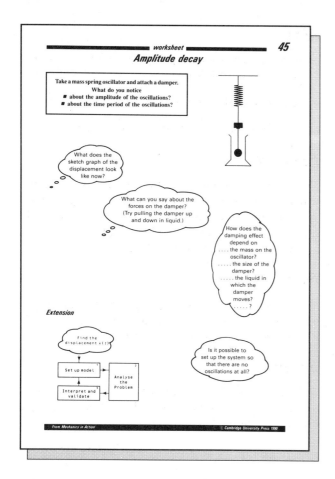

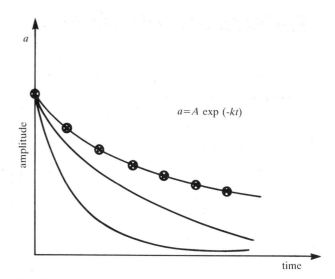

$$a = A \exp(-kt)$$

Fig. 10.8

Similarly the data collected on time period should give T = constant. This hypothesis can further be tested using A-level statistical techniques.

On the other hand, the situation can be taken as a starting point for refining the Newtonian model (simple harmonic motion) to account for amplitude decay. The equation $m\ddot{x} + kx = 0$ is clearly inadequate because of the damping force. Different damping forces might be considered, for example $R \propto \dot{x}$, or $R \propto \dot{x}^2$ (see chapter 2, section 2.3(c)). A solution such as

$$x = a(t) \cos \omega t$$

might be tried as a solution to the refined differential equation

$$m\ddot{x} + c\dot{x} + kx = 0$$

(Clearly if the students have studied second order differential equations then the approach taken in 'Solutions' may be adopted.) This can yield a recognisable differential equation in A and a formula for ω.

Aims

- To use exponential functions to model the amplitude decay of oscillations.
- To refine the simple harmonic motion model to incorporate resisted motion.

Equipment

It is possible to use a spring system without a damper connected, but it will take a long time to collect the data on amplitude decay.

With a damper connected to the mass, such as that provided with the Leeds Mechanics Kit, it will be possible to demonstrate overdamping, critical and sub-critical damping.

Plan

As with much of the practical work, it is possible to use the practical with students who know very little or no mechanics. Without the damper, simply collect data on the amplitude and period of oscillations every 30 seconds for 10 minutes. Ask the students to enter their data on a graph plotter and try to produce a function to fit the data. This can then be used to predict the amplitude decay of the system from any initial amplitude.

Solution

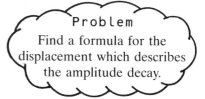

Problem
Find a formula for the displacement which describes the amplitude decay.

Assume the bob is a particle of mass m subject to a Hookean tension, a weight mg and a resistive force of magnitude $c\dot{x}$ as shown in the diagram:

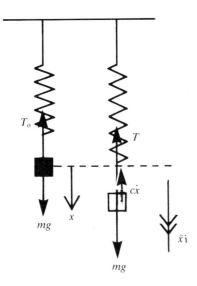

Fig. 10.9

Apply Newton's second law in the direction **i**,

The differential equation is now

$$m\ddot{x} + c\dot{x} + kx = 0.$$

Substituting $x = A\,e^{\omega t}$ gives

$$m\omega^2 + c\omega + k = 0,$$

$$\omega = -\frac{c}{2m} \pm \sqrt{\frac{c^2}{4m^2} - \frac{k}{m}}$$

or

$$\omega_1 = \frac{-c + \sqrt{c^2 - 4mk}}{2m}; \quad \omega_2 = \frac{-c - \sqrt{c^2 - 4mk}}{2m}.$$

The solutions depend on the sign of $c^2 - 4mk$:

(a) if $c^2 > 4mk$ then $x = a\exp(\omega_1 t) + b\exp(\omega_2 t)$,
(b) if $c^2 = 4mk$ then $x = (a + bt)\exp(-ct/2m)$,

(c) if $c^2 < 4mk$ then $\omega = \dfrac{-c}{2m} \pm in$,

where

$$n = \sqrt{\left(\frac{4mk - c^2}{2m}\right)}$$

and therefore

$$x = A\exp(-ct/2m)\cos(nt + \phi)$$

for some arbitrary constant, ϕ.

The condition $c^2 >, =, < 4mk$ clearly depends on the relative magnitudes of the resistive force, the spring stiffness and mass.

1 If $c^2 < 4mk$, $x = a\exp(\omega_1 t) + b\exp(\omega_2 t)$ where ω_1, ω_2 are both real and negative. This corresponds to **overdamping**; no oscillations are observed at all, and the mass returns to equilibrium when displaced.

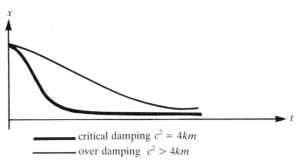

Fig 10.10a

2 If $c^2 = 4mk$, $x = (a + bt)\exp(-ct/2m)$ corresponds to **critical damping**. No oscillations are observed, and the mass returns to equilibrium most rapidly.

3 If $c^2 < 4mk$, $x = A\exp(-ct/2m)\cos(nt + \phi)$ corresponds to **damped** oscillations. The amplitude of the oscillation $A\exp(-ct/2m)$ decays exponentially. The half life of the damping of amplitude is determined by c/m, the resistive force per unit mass. The time period is constant $2\pi/n$. As $c \to 0$ this solution approaches simple harmonic motion:

$$x = A\cos\omega t \quad \text{where} \quad \omega^2 = \frac{k}{m}.$$

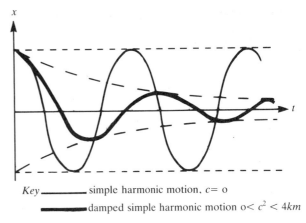

Fig 10.10b

As one would expect, this ensures that if the resistive coefficient is **small** compared with the mass, then simple harmonic motion is a good model for a long time, until t is of the order of m/c in fact.

Resonance

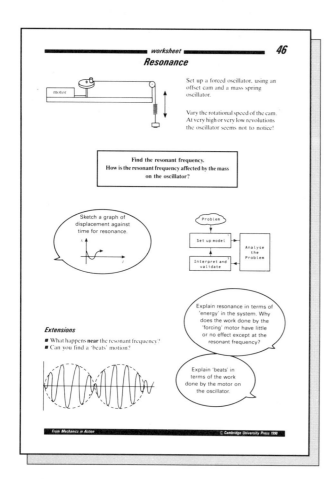

Aims

■ To investigate how the resonant frequency varies with the mass on the spring.

■ To provide a demonstration of resonance which can be modelled using a differential equation.

Equipment

The motor driving an offset cam is shown on the worksheet. Such a system is provided in the Leeds Mechanics Kit.

Plan

The relationship between resonant frequency and mass can be explored using simple empirical techniques.

However, for students who have studied the necessary theory the modelling with differential equations provides a tough problem which will require some help from the teacher. The key idea is to introduce a variable $y = r \sin ft$ for the depth of the top of the spring below its mean position.

Solution

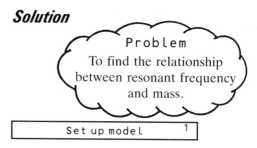

Problem
To find the relationship between resonant frequency and mass.

| Set up model | 1 |

Assume no air resistance and Hooke's law for tension in the spring. The displacement of the top of the spring is $y = r \sin ft$, where r is the radius of the cam and f is its frequency of rotation.

| Analyse the problem | 2 |

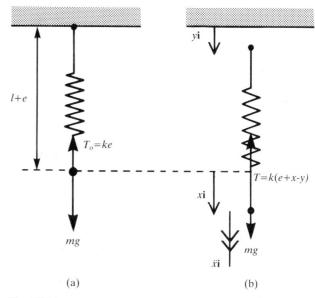

(a)   (b)

Fig. 10.11

The diagrams (Figs. 10.11 (a) and (b)) show the mass in equilibrium and at time t, with displacement x beyond equilibrium. Using Hooke's law

$$T_0 = ke$$

and

$$T = k(e + x - y).$$

Newton's second law gives

$$mg - T_0 = 0 \quad \text{(equilibrium)}$$

and

$$mg - T = m\ddot{x}.$$

Therefore

$$-k(x-y) = m\ddot{x}$$
$$m\ddot{x} + kx = ky$$

or

$$m\ddot{x} + kx = kr \sin ft$$

This can be written in terms of ω, the natural frequency of the mass spring oscillator:

$$\ddot{x} + \omega^2 x = r\omega^2 \sin ft$$

$$\omega^2 = \frac{k}{m}.$$

where

A particular integral is $x = a \sin ft$

where $\qquad a = \dfrac{r\omega^2}{\omega^2 - f^2}$ provided $f \neq \omega$.

If $f = \omega$ then the particular integral is $x = At \cos ft$

where $A = -\dfrac{r\omega^2}{2f}$ or $-\dfrac{r\omega}{2}$.

Interpret and validate [3]

1 When $f \neq \omega$, the displacement of the mass is bounded, consisting in general of a complementary function and a particular integral.

$$x(t) = A \cos \omega t + B \sin \omega t + \frac{r\omega^2}{(\omega^2 - f^2)} \sin ft.$$

2 Resonance occurs when $f = \omega$, which is when f is equal to the natural frequency of oscillation of the spring.

In this case $x = At \sin ft$ describes an oscillation whose amplitude (At) grows linearly with time.

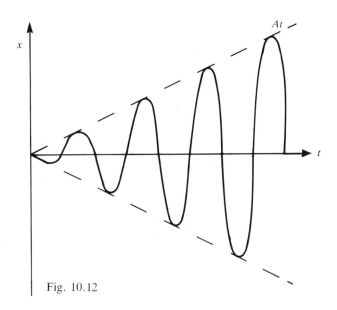

Fig. 10.12

3 The resonant frequency varies as $(\sqrt{m})^{-1}$!

Extensions

1 Incorporating damping terms in the differential equation will show how resonance is **eventually** controlled. A question of real importance is 'How much damping needs to be introduced to control it?'.
2 Attaching two mass spring systems in series will nullify resonance. Why?
3 Investigate what happens when f is close but not equal to ω and the phenomenon of 'beats' arises.

The simple pendulum

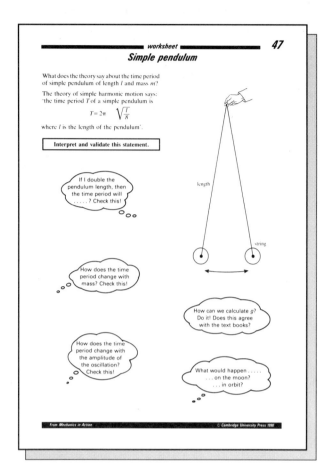

Aims

- To validate the theory of the simple pendulum in practice.
- To provide experience of the interpretation and validation of the solution to a well known problem.
- To illustrate the need to bear in mind the assumptions on which the model is based.

Equipment

A length of string, several 100 g and 10 g masses and a stopwatch.

Plan

After having studied the theory of the simple pendulum, students should be encouraged to validate as many aspects of the theory as possible. This is a group activity. Students may need to be prompted to consider the importance of the following assumptions on which the theory is based:

- the mass is a particle.
- the string is light and inextensible.
- the point of support is **fixed**.
- the amplitude of oscillation is 'small enough' so that $\sin \theta \simeq \theta$ in radians.

The solution of this problem and its interpretation and validation is dealt with in chapter 2.

The main points to get out of the formula

$$T = 2\pi \sqrt{\frac{l}{g}}$$ are as follows.

1 The time period is proportional to the square root of the length of the string.

2 The time period is independent of the mass (testing this hypothesis may involve A-level statistics).

3 Experimental results can be used to calculate g.

4 Numerical predictions can be made for time periods such as

when $l = 3$ m, $T = 3.4$ s.

5 The independence of the mass in **2** assumes that the mass is sufficient to cause oscillations, and is relatively large compared to the weight of the string. The solution becomes invalid when m is small, or zero.

6 The solution is inaccurate for amplitudes over about 45°. It is **very** accurate for amplitudes up to about 15°.

7 The solution is inaccurate if the length of the string is small compared to the size of the mass attached, even if l is taken to be the distance from the point of attachment of the string to the centre of the mass. If the size of the mass is not insignificant, a refined compound pendulum model may be required.

8 Over a period of 24 hours, the plane of oscillation will appear to rotate about the vertical.

Extensions

The problem can lead to a study of compound pendulums (see the worksheet 'The compound pendulum').

The compound pendulum

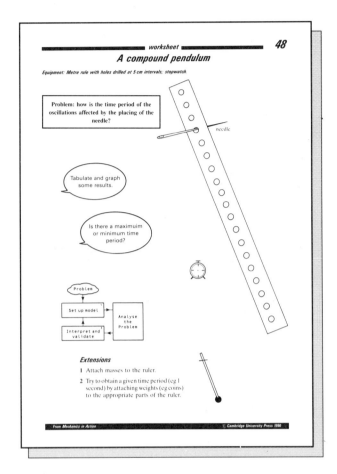

Aim

- To provide a simple practical problem to investigate empirically or to model with rigid body dynamics.

Equipment

See the worksheet.

Plan

After studying rigid body dynamics, present the problem to groups as a modelling investigation or as one of a number of such investigations.

Solution

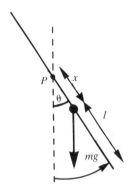

Problem
How is the time period affected by the position of the needle?

| Set up model | 1 |

Assume there is no torque on the ruler from the needle; neglect air resistance and assume the oscillations are, as usual, 'small'. Assume the ruler to be a uniform rod of length $2l$ and mass m, so its moment of inertia about P is

$$mx^2 + \frac{ml^2}{3}.$$

| Analyse the problem | 2 |

The torque about the horizontal axis through P is given by Torque = (moment of intertia about P) $\times \ddot{\theta}$ in an anticlockwise sense. Therefore

$$- mg\, x \sin\theta = (mx^2 + \tfrac{1}{3}ml^2)\, \ddot{\theta}$$

and

$$\ddot{\theta} = -\left(\frac{gx}{x^2 + \tfrac{1}{3}l^2}\right)\sin\theta$$

or

$$\ddot{\theta} + \omega^2\theta = 0$$

where

$$\omega = \sqrt{\frac{gx}{x^2 + \tfrac{1}{3}l^2}}.$$

This gives a solution for time period

$$T = 2\pi \sqrt{\frac{x^2 + \tfrac{1}{3}l^2}{gx}}.$$

153

1 A graph of T against x is shown in Fig. 10.13

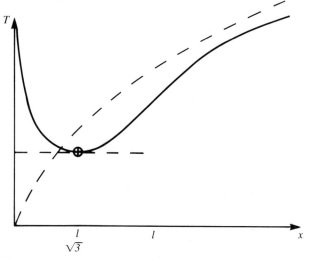

Fig. 10.13

2 The turning point is at $x = \dfrac{l}{\sqrt{3}}$. Therefore for a metre ruler, the minimum time period is obtained when the needle is 29 cm from the centre.

3 This minimum time period is $2\pi \sqrt{\dfrac{2l}{g\sqrt{3}}} \simeq 1.5$ seconds.

4 The time period can be rewritten as

$$T = 2\sqrt{\frac{x}{g}}\sqrt{1 + \frac{l^2}{3x^2}},$$

showing that the time period is always greater than $2\pi\sqrt{\dfrac{x}{g}}$, the time for a simple pendulum of length x.

But if x is large compared with l, then $T \simeq 2\pi\sqrt{\dfrac{x}{g}}$ recovers the simple pendulum solution, as might be expected. This is shown by the dotted curve in fig. 10.13.

11

Impact

This chapter contains activities which can be used in a variety of ways.

Bouncing ball (1) and (2)   can be used with pure mathematics classes to provide data which can be modelled using linear, quadratic, square root or reciprocal functions, geometric series and exponential functions.

Bouncing ball (1)   can also be used to introduce mechanics classes to the law of restitution.

Bouncing ball (2)   can also be used to provide a number of modelling investigations and is therefore an ideal starting point for A-level coursework.

Newton's cradle   provides a practical and a modelling investigation which can be used to introduce or apply the theory for impact with mechanics classes.

Rebound   can provide a modelling investigation of the rebound of a snooker ball off a cushion. A number of purely geometrical and mechanics extensions are suggested which can lead to an extended project.

The superball as a deadly weapon   provides a real and challenging problem to solve with the theory for impact. This is suitable for upper sixth mechanics students.

Bouncing ball (1) and (2)

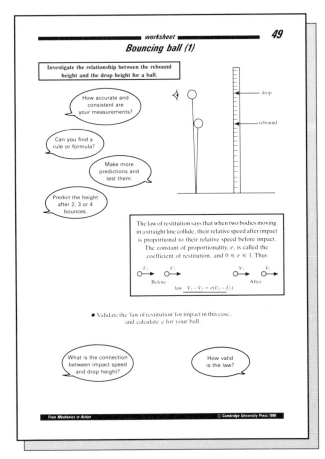

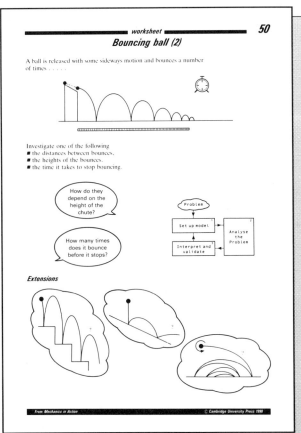

Aims

- To provide modelling problems with an emphasis on pure mathematics: the fitting of functions to data, the use of geometric progressions.
- To provide experimental validation of the law of restitution.
- To provide modelling problems involving dynamics and the law of restitution.

Equipment

(1) Rulers and superballs.
(2) Rulers, superballs, stop watch, sugar paper and felt tip pens. A chute made from a cereal packet and cardboard tube.

Plan

Different plans are appropriate for each of the two worksheets and for students at different stages.

Plan A

For a lower sixth pure mathematics class with little or no knowledge of mechanics; duration 60–70 minutes.

Split the class into groups of three and provide each with a metre ruler, superball and worksheet (1). A variety of other balls should also be available; ping-pong, tennis and blu-tak balls.

Allow them 40 minutes to record results and find a graphical relationship between drop height and rebound height. You may:

(a) encourage them to indicate an estimated error or an intuitive confidence interval for drop and rebound heights, for example $h = 69$ cm ± 3 cm.
(b) examine the **consistency** of their rebound heights; it may help them to use a flat and even floor surface.
(c) prompt them to produce a formula. This will encourage them to introduce variables.

After about 40 minutes, it is a good idea to ask each group to report to the class, compare results and **interpret** them.

- If one graph is steeper than another then what does this mean?
- If a graph does not go through the origin what does this mean? (It may well be that drop and rebound heights were not measured from the bottom of the ball!)
- If the graph 'curves' for a ping-pong ball, then what can this be due to?

Encourage students to validate by interpolating their graphs or substituting new values into their formulae. A good question concerns the extent to which they can **extrapolate** using their model.

- If you drop the ball from 100 metres, how far will it rebound? Are you confident that this is a valid prediction?

The use of a ping-pong ball or a practice golf ball (with holes in it) will suggest the importance of air resistance in modifying the model for larger drop heights. If there is time, you may discuss the implication of Newton's law of restitution. The result that (rebound speed) = $e \times$ (impact speed) can then be linked to their variables by:

$$\text{height fallen} \propto (\text{impact speed})^2$$

so that

$$\text{rebound height} = e^2 (\text{drop height}).$$

This can then be compared with their formulae or graphical results.

If the students are already familiar with and understand the law of restitution then an alternative presentation will be 'validate the law for a bouncing ball'.

Plan B

An A-level pure or applied mathematics class investigation, 'Modelling with Functions'. Duration 70 minutes plus individual work.

Each group will require a graphic calculator or access to a function graph plotter.

Introduction (15 minutes)

Although worksheet (2) can stand alone, we suggest you introduce the activity yourself. Set up a long, smooth table with several sheets of sugar paper upon it. Fix a chute on a pile of books; this allows you to repeat the experiment always giving the ball the same sideways speed at take off.

Roll the ball down the chute and check that you get plenty of bounces on the sugar paper. Now wet the ball from a cup of water and repeat. The ball should leave marks where it bounces. If not, use a heavier ball or more water. It may help to sprinkle chalk dust or talc and smooth down with a board duster. Draw dots on the bounce marks with a felt tip pen.

'The problem is to investigate the distances between bounces – can you find a formula or rule?'

You may wish to construct an instant graph by transferring the bounce marks onto a long strip of paper: just cut the paper at the marks and paste them side by side onto sugar paper.

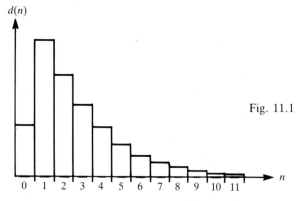

Fig. 11.1

Group work (55 minutes)

Each group should be given sugar paper, a superball, materials for making a chute and water. (Try to avoid balls with seams since these produce occasional erratic bounces!) Students should produce their own experimental results and try to find a formula to fit their data. Several functions may be considered depending on the prior knowledge of the class.

$$d(n) = a - Kn,$$

$$d(n) = \frac{a}{b+n},$$

$$d(n) = aK^n,$$

$$d(n) = ae^{-kn}.$$

If you intervene and curtail the exploration of the different functions then valuable teaching opportunities may be lost. In particular:

■ that the linear model is **locally good**, but eventually becomes invalid for large n, and particularly when $d(n)$ becomes negative.

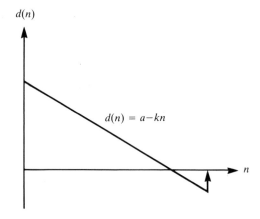

Fig. 11.2

■ that a local linear approximation is related to the **derivative** of the 'actual function'.
■ that over limited ranges of n, different functions are equally valid models and there are other reasons for preferring one model to another. (In this case, perhaps as a result of the lesson in plan A concerning the ratios of successive bounces.)
■ that aK^n and ae^{-kn} are identical functions for a suitable choice of k and K (in fact $-k = \log K$).

Note that the modelling of the data collected on a function graph plotter or graphic calculator involves students in entering data and varying parameters until the functions fit the data.

It is crucial that the activity does not stop at this stage. Encourage the group to use their formulae to make predictions and test them.

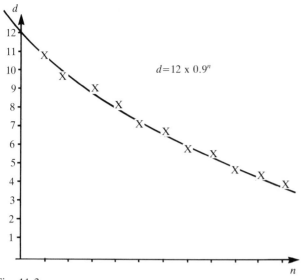

Fig. 11.3

- What happens as $n \to \infty$?
- What can you say about the total distance travelled before bouncing stops?
- What could you say about a ball whose first three bounces were 1 m and 0.6 m apart?

Plan C

An A-level mechanics class, applying known theory to bouncing ball problems.

Introduction (10 minutes)

As a class, obtain a value for the coefficient of restitution of a ball. It is assumed here that the class have already met the law of restitution and the principle of conservation of energy

$$mgh = \tfrac{1}{2}mv^2$$

also

$$mgr = \tfrac{1}{2}m\,(ev)^2$$

and so

$$h = e^2r.$$

Now introduce the problem with a challenge perhaps!

'For £500, I want you to calculate, in the next 60 minutes, how long it will take for the ball to stop bouncing.'

Group work (60 minutes)

Each group is given a ball, metre ruler and stopwatch. They should solve the problem for **their** ball, validate **their** theory and finally generalise to give a result for **your** ball.

You may conclude the activity when you wish (or next lesson) by writing down the results of each group and testing them with the ball and several stopwatches. An argument can then ensue as to who collects the £500!

- Did I drop the ball from **exactly** 1 m?
- Which bit of the floor did you use?
- Some of the stopwatch readings were close to my result! (Use a digital stopwatch with a six figure reading: no-one will get the **exact** time!) Your £500 is safe!

Solutions

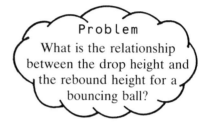

Problem
What is the relationship between the drop height and the rebound height for a bouncing ball?

| Set up model | 1 |

Assume the ball is a particle, released from a height h, and rebounding to a height r. Assume air resistance is negligible.

Assume that Newton's law of restitution is valid with e representing the ball's coefficient of restitution.

| Analyse the problem | 2 |

If the impact speed is v then the law of conservation of energy gives

$$mgh = \tfrac{1}{2}mv^2. \qquad (11.1)$$

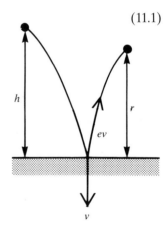

The rebound speed is ev, and therefore using the conservation of energy again

$$mgr = \tfrac{1}{2}m\,(ev)^2. \qquad (11.2)$$

Eliminating v from (11.1) and (11.2) gives

$$r = e^2h.$$

The drop height and rebound height are directly proportional and give a straight line graph of gradient e^2.

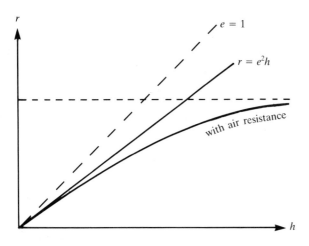

Fig. 11.4

This solution assumes that air resistance is negligible. When a ball is dropped from a 'large height' air resistance becomes significant and the impact speed approaches a terminal speed which is independent of the drop height. The graph will curve away from the straight line $r = e^2h$ and approach an asymptote $r = $ constant.

The assumption that the ball is a **particle** means that consideration needs to be given to where precisely r and h are measured from: in fact, measurements should be made from the bottom of the ball to the floor. For large balls this is clearly an important factor to get right.

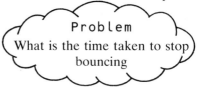

> **Problem**
> What is the height of the ball after n bounces $h(n)$?

The previous model and analysis applies to give

$$h(n) = e^2h(n-1)$$

and so

$$h(n) = e^{2n}h(0)$$
$$h(n) = e^{2n}h.$$

This result can be used to predict:

(a) the height after 5 bounces, say.
(b) the number of bounces occurring before the height is halved, say. Using logarithms

$$\log h(n) = 2n \log e + \log h.$$

Therefore

$$\log \tfrac{1}{2} = 2n \log e$$

which implies

$$n = \frac{\log \frac{1}{2}}{2 \log e}.$$

(c) that there are an infinite number of bounces before $h(n) = 0$, i.e. before the ball stops bouncing!

> **Problem**
> What is the time taken to stop bouncing

With the same model as above, the time taken to fall from a given height h is given by

$$h = \tfrac{1}{2}gt^2$$

and so the time taken to fall from $h(n)$ is

$$t(n) = \sqrt{\frac{2h(n)}{g}}$$

and the time between successive bounces is $2t(n)$.

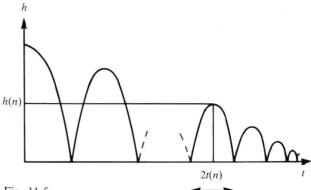

Fig. 11.5

The total time T to stop bouncing is the sum of the infinite series

$$T = t(0) + 2t(1) + 2t(2) + \ldots$$

$$= \sqrt{\frac{2h}{g}} + 2\sqrt{\frac{2h(1)}{g}} + 2\sqrt{\frac{2h(2)}{g}} + \ldots$$

since

$$h(n) = e^{2n}h$$

therefore

$$t(n) = e^n t(0)$$

and

$$T = \sqrt{\frac{2h}{g}}[1 + 2e + 2e^2 + 2e^3 + \ldots]$$

Therefore

$$T = \sqrt{\frac{2h}{g}}[1 + 2e(1 + e + e^2 + \ldots)]$$

$$T = \sqrt{\frac{2h}{g}}\left[1 + \frac{2e}{1-e}\right] = \sqrt{\frac{2h}{g}}\frac{(1+e)}{(1-e)}$$

| Interpret and validate 3 |

This result means that:

(a) the ball does indeed stop bouncing in a finite time which can be calculated given e, h and g.
(b) the time T is proportional to the square root of the initial drop height.

These results should be validated. Note, however, that if e is close to 1, small errors in e can lead to large errors in total time, T. Compare predictions for T when $e = 0.96$ and 0.98 for example. This could lead to a discussion of analysis of errors and then using the relation

$$\delta T \simeq \frac{dT}{de} \times \delta e$$

obtain the result

$$\delta T \simeq \sqrt{\frac{2h}{g}}\frac{[(1-e) + (1+e)]}{(1-e)^2}\delta e$$

$$\delta T \simeq \sqrt{\frac{2h}{g}}\frac{e}{(1-e)^2}\delta e \to \infty \text{ as } e \to 1!$$

Problem

Find the distance between successive bounces.

| Set up model   1 |

In addition to the previous assumptions, assume that the ball is given an initial sideways speed V but no spin, so that V is assumed to be constant.

| Analyse the problem   2 |

Let $d(n)$ be the distance between the nth and the $(n+1)$th bounces. In the previous analysis we obtained

$$t(n) = e^n\sqrt{\frac{2h}{g}} = e^n t(0) = et(n-1).$$

Therefore

$$d(n) = Vt(n) = eVt(n-1)$$

$$= ed(n-1)$$

giving

$$d(n+1) = e^n d(1)$$

for $n = 0, 1, 2, \ldots$

This result can be tested against real data in which $d(1)$ is the distance between the first two bounces for a ball with coefficient of restitution e.

The distance from take off to the first bounce is not modelled by this formula. We can consider this distance to be a part of one bounce.

$d(0)$ is only a part of a bounce and is not described by the formula.

The above formula can be used to find the cumulative distance travelled before rolling begins.

$$D = d(0) + \sum_0^\infty d(n+1) = d(0) + \sum_0^\infty e^n d(1)$$

$$D = d(0) + \frac{d(1)}{1-e}.$$

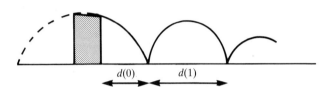

Fig. 11.6

😟 *Misconceptions*

Confusion can arise if students fail to appreciate:

(a) the true status of Newton's law of restitution; that it is an empirical or 'modelling' law (see chapter 2).
(b) that an infinite series can arise and give a finite sum **in practice**. We may argue that only a finite number of observable bounces do occur or that beyond a certain point the bounces are of the same order of magnitude as the vibrations of the ball on the floor.

Extensions

A number of extensions are suggested on worksheet (2). Note the difference between bouncing downstairs, which can theoretically continue forever if the step size is appropriate, and bouncing down a slope which must stop in a finite time!

Newton's cradle

Aims

■ To provide a practical investigation using the theory for impact.

Equipment

One Newton's cradle. Provide groups with ball bearings to simulate the cradle.

Plan

Introduction

Issue the worksheet to groups. Ask them to guess **first**. They may know very little mechanics and guess intuitively, or they may be encouraged to apply laws of impact to predict what happens.

Practical

Provide them with a cradle or get them to simulate by rolling ball bearings or snooker balls between rulers as shown in Fig. 11.7.

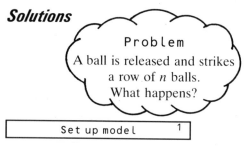

Fig. 11.7

Ask them to validate their predictions.

Solutions

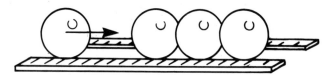

Problem

A ball is released and strikes a row of n balls. What happens?

```
Set up model        1
```

Assume the balls are identical, perfectly elastic, particles of mass m. Assume the particles are constrained to move in a line on a smooth plane and there is a **small gap** between the stationary particles. This is equivalent to assuming that the impact can be modelled as a sequence of distinct collisions each involving two particles.

```
Analyse the problem   2
```

Consider the first impact.
   Let the speeds before and after impact be as shown.

Before impact

After impact

Conservation of momentum gives

$$mV_1 + mV_2 = mU.$$

A collision involving perfectly elastic particles implies conservation of kinetic energy

$$\tfrac{1}{2}mV_1^2 + \tfrac{1}{2}mV_2^2 = \tfrac{1}{2}mU^2.$$

Therefore

$$(U - V_2)^2 + V_2^2 = U^2$$
$$V_2 (V_2 - U) = 0.$$

This implies

$$V_2 = 0, \quad V_1 = U$$

or

$$V_2 = \ U, V_1 = 0$$

and the latter represents the distribution of velocities after impact. By induction all subsequent impacts are similar.

```
Interpret and validate  3
```

It follows that the end ball will move off while the others remain stationary. Given enough balls the theory breaks down since in practice, a small amount of energy is lost with each impact. This is also observed in Newton's cradle as the collisions become 'less pronounced' and eventually cease.

Extension

By considering sequences of impacts all the other problems can be solved similarly. For example for two impacting balls:

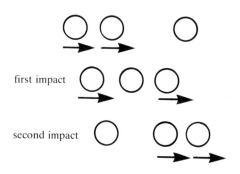

first impact

second impact

Rebound

Aims

■ To provide practice in modelling a practical problem:

  (a) empirically,
  (b) using the laws of impact.

Equipment

If possible, a group might get access to a pool or snooker table. Otherwise use a ball on an even floor striking a wall. Talc or chalk dust can help to leave a trace of the ball's path.

Plan

This is intended to be an activity which students can manage without introduction.

The graphical work should lead to searching for a formula with the aid of a graph plotter or graphic calculator.

If students have studied the laws of impact, they should be encouraged to model the situation using a coefficient of restitution $e < 1$.

Note that results obtained from a function graph plotter are extremely unlikely to be the same as those obtained from the laws of impact ($\tan \phi = e \tan \theta$)!

Solution

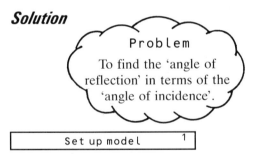

Problem

To find the 'angle of reflection' in terms of the 'angle of incidence'.

| Set up model | 1 |

Let the angles of incidence and reflection be θ and ϕ, and the corresponding speeds be U and V.

Assume the ball is a particle of mass m, striking a smooth inelastic plane with coefficient of restitution e.

| Analyse the problem | 2 |

Since the impulse of the collision is normal to the plane, the component of momentum parallel to the plane is conserved (see Fig. 11.8)

$$mU \cos \theta = mV \cos \phi. \qquad (11.3)$$

The law of restitution gives

$$V \sin \phi = eU \sin \theta \qquad (11.4)$$

so that eliminating U and V gives

$$\tan \phi = e \tan \theta.$$

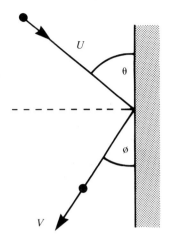

Fig. 11.8

| Interpret and validate | 3 |

Values of θ and ϕ can be predicted and tested to validate this result. For example

$$\theta = 0 \quad \text{implies } \phi = 0.$$

Graphs of θ against ϕ for different e are shown in Fig. 11.9.

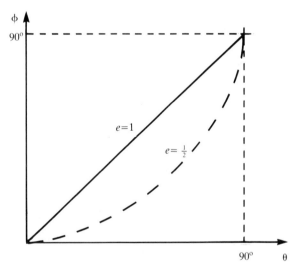

Fig. 11.9

Clearly $e = 1$ gives $\phi = \theta$ and the nearer e is to 1 the more accurately the reflection property holds.

If the ball is given spin this result is not valid, since the assumption that the impulse is normal to the plane fails. In fact there is a tangential component due to friction of impulse when a spinning ball strikes a real cushion.

Extensions

There are a large number of good 'geometrical' modelling problems associated with snooker which require only circle geometry and calculus.

1 Estimate the angle within which a ball must be cued to go into a pocket.

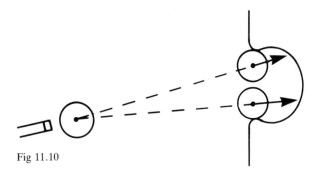

Fig 11.10

2 Which shots are easier than others? For example, compare (a) and (b).

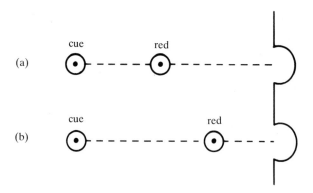

Fig. 11.11

3 Why should you make your 'bridge' as near the cue ball as possible?

In addition there are **mechanics** modelling problems which involve the laws of impact for oblique collisions and take account of the spin of the snooker ball.

4 Why do the cue ball and red ball always move at right angles to each other after impact?

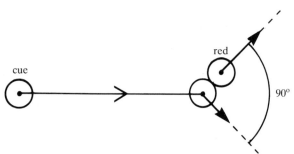

Fig. 11.12

5 Why is the cushion height $\frac{2}{3}$ of the diameter of the balls?
6 Why does the cue ball 'roll on' after impact, even when it strikes a red head-on?

The superball as a deadly weapon

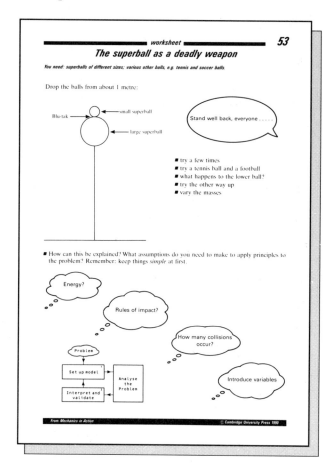

Aims

- To provide a real problem to model with the laws of impact.
- To appreciate the implications of making a 'crucial assumption'.

Equipment

See the worksheet.

Plan

Practise this trick yourself until you can do it with superballs or with a tennis ball and football. You need to release the large ball without giving it rotational speed, so that when it hits the ground the small ball is still on top of the large ball.

Then let the class attack the problem in groups.

Encourage them to:

(a) explore the practical.
(b) make drastic assumptions, for example, there is no energy loss ($e = 1$).

The key assumption is that the impact can be modelled as in Newton's cradle as two **separate** impacts; first of the large ball with the floor, then of the large ball with the small one. You should be prepared to help if you see frustration becoming overwhelming, but this problem **can** motivate students to struggle for a long time without help.

Solution

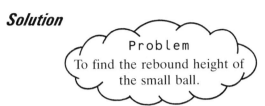

Problem
To find the rebound height of the small ball.

Set up model   1

The balls are modelled as particles of mass m and km, released from a height of 1 metre. The collisions are separate, perfectly elastic and there is no energy loss due to air resistance.

Analyse the problem   2

The balls reach the ground with speed V, where $V^2 = 2g \times 1$,

$$V = \sqrt{2g}.$$

The large ball rebounds with speed V, and subsequently impacts with the small ball, giving speeds after the second impact of V_1 and V_2 as shown.

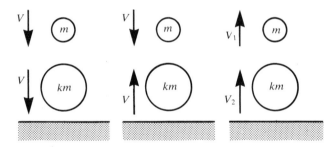

Fig. 11A

Conservation of momentum gives

$$mV_1 + kmV_2 = kmV - mV$$

or

$$V_1 + kV_2 = (k - 1)V.$$

The law of restitution with $e = 1$ gives

$$V_1 - V_2 = 2V.$$

Note that we could equally use conservation of kinetic energy, when $e = 1$.

Eliminating V_2 gives $V_1 = \dfrac{(3k-1)V}{(k+1)}$ or $\dfrac{(3k-1)}{k+1}\sqrt{2g}.$

The final height h, of the small ball will therefore be

$$h = \frac{V_1^2}{2g} = \left(\frac{3k-1}{k+1}\right)^2.$$

Interpret and validate ³

1 If we know the ratio of the masses of the balls, k, we can calculate the height.
For example

$$k = 2, \quad h = \frac{25}{9} = 2.8 \text{ metres}.$$

These calculated values are very high, because we assume no energy loss.

2 As $k \to \infty$, h achieves a maximum of 9 metres. The intuition that there is no limit to the possible height achieved is proved wrong, even with no energy loss.

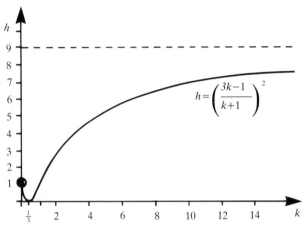

Fig. 11.13

Extensions

A refined model would assume coefficients of restitution e_1 and e_2 at the impacts, and obtain

$$h = \left(\frac{(e_1 + e_2 + e_1 e_2)\,k - 1}{k + 1}\right)^2.$$

Furthermore, the height of rebound of the **large** ball can be calculated to give

$$\left(\frac{e_1 k - (1 + e_2 + e_1 e_2)}{1 + k}\right)^2.$$

which reduces to

$$\left(\frac{k-3}{k+1}\right)^2.$$

when $e_1 = e_2 = 1$,

Further interpretations and validation can therefore be carried out with specified values for e_1, e_2 and k.

165

REFERENCES

(1) Kepler, J., *Astronomia Nova* (Commentaries on the Motion of Mars) (1609), Johannes Kepler, Gesammelte Werke, Vol. 3, C.H. Beck, Munich, 1937.

(2) Kepler, J., *The Harmonies of the World* (1619) (trans. C.G. Wallis), Encyclopaedia Britannica, Inc., Great Books, Vol. 16, pp. 1009-85 Chicago, 1952.

(3) Roller, D. H., *The De Magnete of William Gilbert*, Menno Hertzberger, Amsterdam, 1959.

(4) Newton, I., *Principia* (Mathematical Principles of Natural Philosophy), (1687). F. Cajori's revision of A. Motte's translation (1729), University of California Press, Berkeley, California, 1960.

(5) Eves, H., *An Introduction to the History of Mathematics*, Saunders College Publishing, New York, 1983.

(6) Shah, I., *Caravan of Dreams*, Octagon Press, London, 1968.

(7) Savage, M. D., and Williams, J. S., 'Centrifugal Force: Fact or Fiction?' *Phys. Educ.* **24**, 1989.

(8) Davies, A. J., 'A Simple Friction Modelling Exercise', *Teaching Mathematics and its Applications*, Vol. 6, 1987.

Bathroom scales and a broom

How can I lose weight?

No! Down on the floor

Press up on the ceiling

..... press down on the scales ..... on another set of scales ....

Guess first.
Then explain and draw a force diagram.

Problem 1
Explain why the reading goes down when I press on the floor with a broom.

Problem 2
If I press up on the ceiling, why does the reading go up?

Problem 3
What happens to the scale reading if I stand on the scales while taking the lift from the tenth to the ground floor?

Make up and solve some problems of your own

Modelling with force diagrams

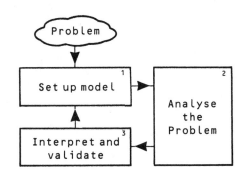

> **Problem**
> Explain why the reading goes down when I press on the floor with a brush.

Set up model 1

Draw a diagram

List assumptions

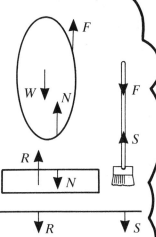

- each body is a particle in equilibrium
- the scales and brush have zero mass

Introduce variables and formulate problem

eg weight W, interactions N, F, S, R as shown, N being the scales' reading.
Problem: Show that N is less than W.

Analyse the problem 2

Apply Newton's laws and write down equations

- NL3:
  insert pairs N, F, S and R in opposite directions on diagram

- NL1:
  for my body: $N + F - W = 0$
  for the scales: $R - N = 0$
  for the brush: $S - F = 0$

Solve the equations

Thus $N = W - F$
and $F = S \geq 0$
so
$$N \leq W.$$

Further
$R + S = N + F = W.$

Interpret and validate 3

Assuming we are pushing with the brush, $S \geq 0$ so $N \leq W$, the reading is less than or equal to the weight. $N = W - F$ means by increasing F we can reduce N, so the harder we push down the less the reading. But $F + N = W$ suggests $F \leq W$ so $S \leq W$. this means we can only push down with force W at most, as long as we are in equilibrium.

For each of your problems try to follow the same procedure as this in your solutions.

Equilibrium

Which of these are in equilibrium?

1 Two 'equal' masses

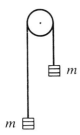

2 Two 'unequal' masses
$M = m$ + a blob of Blu-tak.

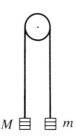

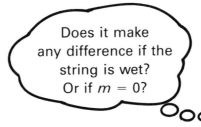

3 As in 1 above, but set the system moving.
Is it still "equilibrium"?

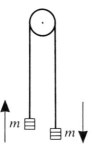

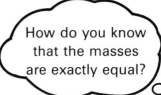

Lines of force

Make notes of what you notice or find out

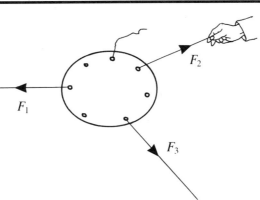

Is it possible to . . . . .

■ . . . . . get the three forces meeting in a point outside the ring/lamina?

■ . . . . . get the three forces to be parallel?

■ . . . . .get the three forces to be non-coplanar?

Extensions

Spring balances (1)

**Guess first, then test,
then explain results**

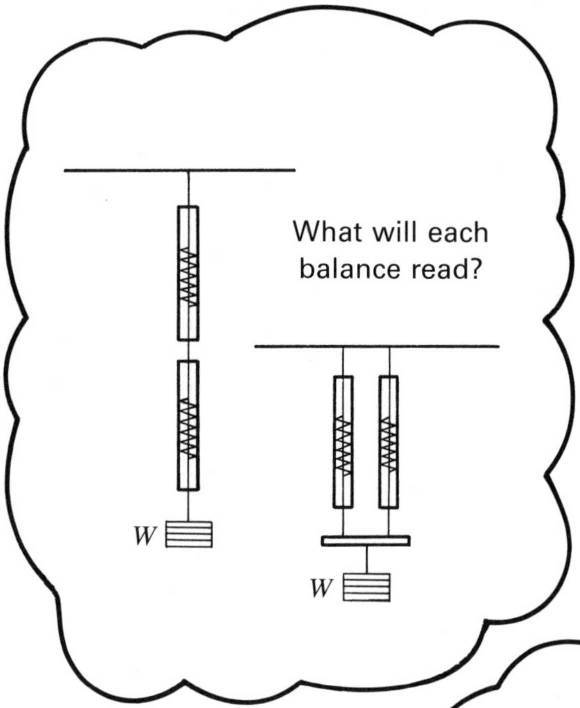

Pull ← Pull

What will each
balance read?

W

W

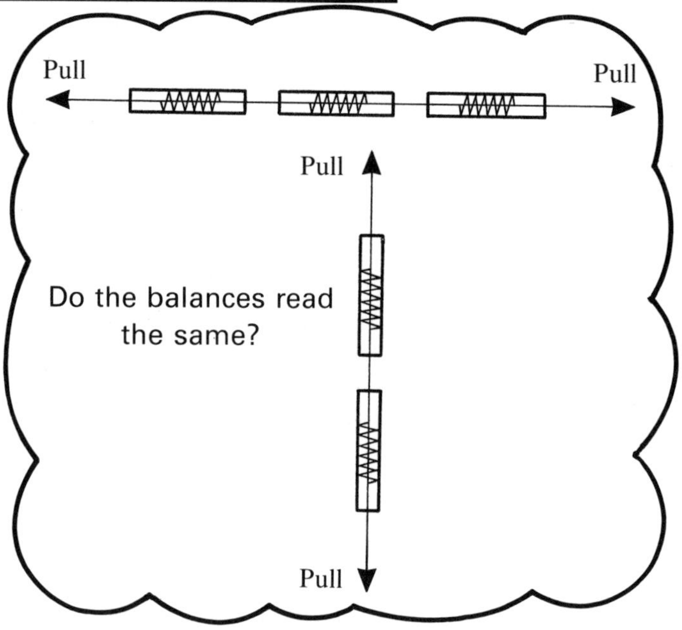

Do the balances read
the same?

Pull

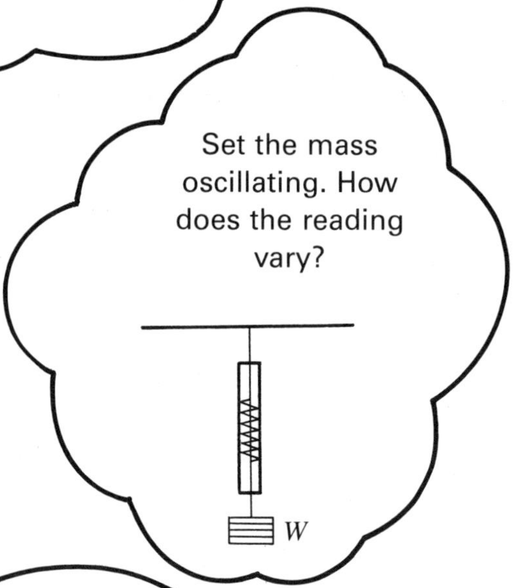

Pull

Set the mass
oscillating. How
does the reading
vary?

W

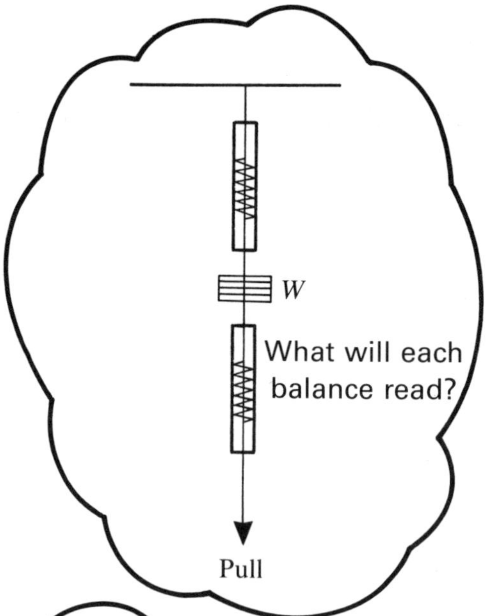

W

What will each
balance read?

Pull

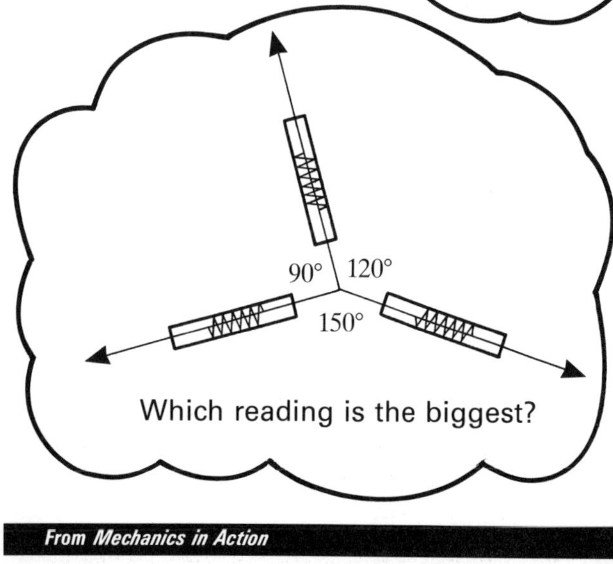

90° 120°

150°

Which reading is the biggest?

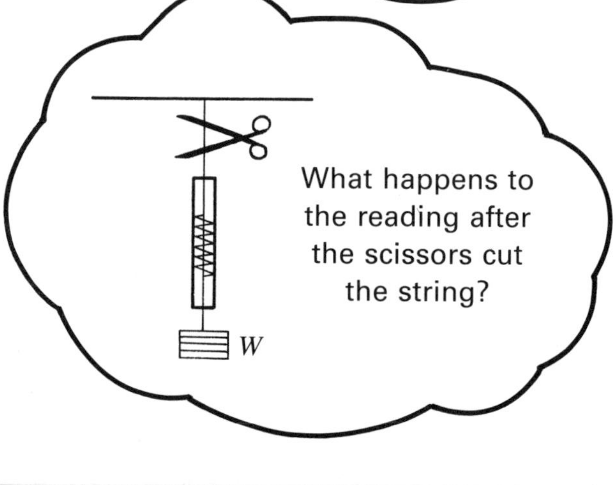

What happens to
the reading after
the scissors cut
the string?

W

Spring balances (2)

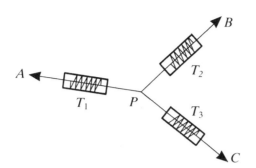

What happens if . . . . .

$T_1 = T_2 = T_3$?

$T_1 + T_2 < T_3$?

APC is a straight line?

$B\hat{P}C = 90°$

$T_1 = T_2 \neq T_3$?
(gradually increase T_3 from zero)

$T_1{}^2 + T_2{}^2 = T_3{}^2$?

> **Record results and write down some ideas
> or rules about forces in equilibrium**

Validate the rule:

Given any force F and any angle θ,
it is possible to resolve F into two
forces, $F\cos\theta$ and $F\sin\theta$.

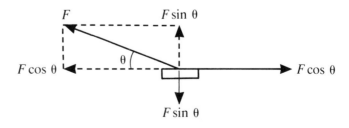

Three masses

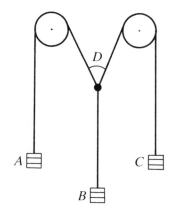

> **Problem**
> What is the connection between masses A, B, C and angle D?

Start with $A = C$
and vary B.

> Describe what happens as you increase B or decrease B.

> Why have $A = C$?
> Why not try $A = B = C$?
> Or why not fix angle D, say 90 degrees, and try to get values of A, B and C?

> Try any values of A, B, C.
> Guess what will happen if . . .
> – you increase A
> – you increase B
> – you increase C
> Were you right?

> Suppose
> $B > A + C$?
> $C > A + B$?
> $A > B + C$?

Parallelogram law

The parallelogram law states that the action of two forces $F_1 \, F_2$ is equivalent to that of a single resultant force **R** which is the vector sum of F_1 and F_2, $\mathbf{R} = \mathbf{F}_1 + \mathbf{F}_2$. **R** is found by constructing the diagonal of the parallelogram as follows:

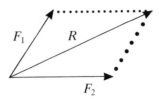

Use the apparatus to validate this law

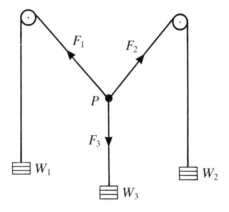

Note that since P is in equilibrium, $\mathbf{F}_3$ balances the resultant of $\mathbf{F}_1$ and $\mathbf{F}_2$

How do you find the magnitude and direction of $\mathbf{F}_1$?

What assumptions must we make about the apparatus? How accurate can we expect our drawing and measurement to be?

Balancing a ruler

1 A ruler will usually balance at its midpoint:

Check that this is so.

2 Attach a 100 g mass to the end of the ruler. Where does it balance now?

100 g

←— *d* —→

Explain your finding.

3 Can you predict where it will balance with 200 g, 300 g? Keep taking results until you have a theory . . . check your theory with predictions.

Generalise your theory or formulae . . . for **any** added mass

. . . for any type of ruler

. . . for a number of rulers stuck together

. . . for various masses attached at various points

Extensions

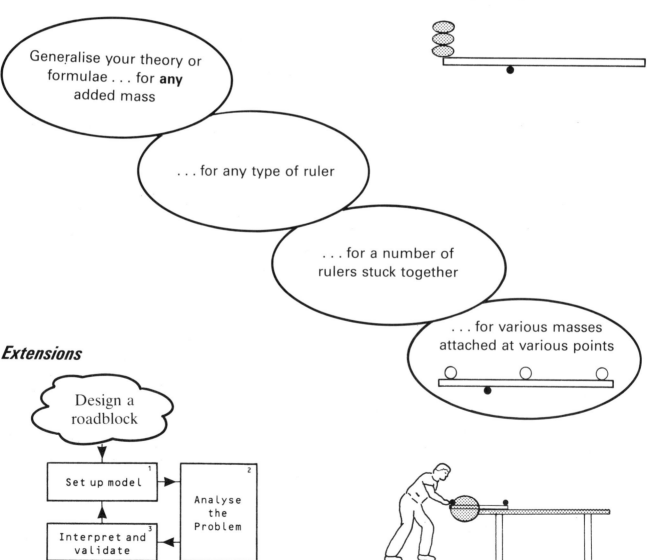

Design a roadblock

Set up model¹	→	Analyse the Problem²
↑↓		
Interpret and validate³	←	

From Mechanics in Action

The law of moments

Find out what the 'law of moments' says. Your task
is to verify the law in the following situations:

Investigation 1

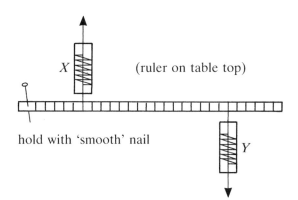

X (ruler on table top)

hold with 'smooth' nail

Y

Investigation 2

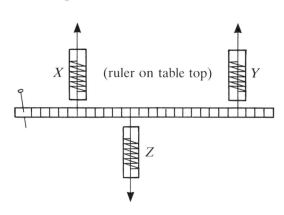

X (ruler on table top) Y

Z

Investigation 3

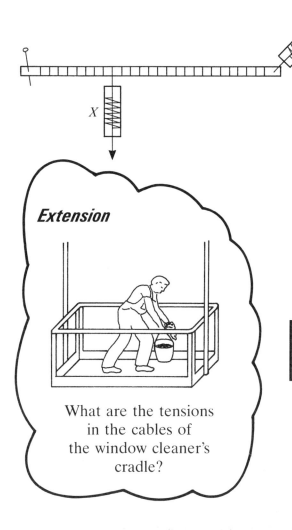

Y

X

Extension

What are the tensions
in the cables of
the window cleaner's
cradle?

Investigation 4

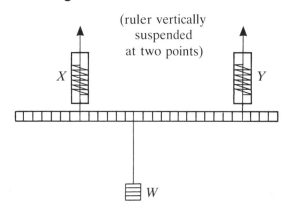

(ruler vertically
suspended
at two points)

X Y

W

Take readings and check that the law of moments
applies within experimental error.

Beam balance

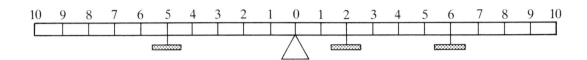

1 You are given 40 g, 70 g and 90 g masses to hang on the beam.

■ Can you do this and make the beam balance?

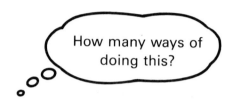

How many ways of doing this?

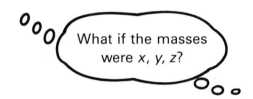

What if the masses were *x, y, z*?

2 You are provided with a heavy block and a light blob of plasticine of unknown masses. You also have five 10 g masses.

■ Can you find the unknown masses? How accurate can you be?

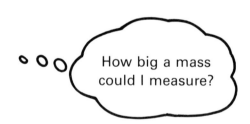

How big a mass could I measure?

3 Here are some puzzles to solve.

■ Place two 10 g masses at adjacent numbers on one side and balance them with three 10 g masses on the other side.

■ Place 10 g masses at *x* and *y* on one side and at 3*x* and $\frac{y}{2}$ on the other so that they balance.

4 Can you invent some good puzzles for the class to solve?

Can you find the mass of the beam?

Over hang

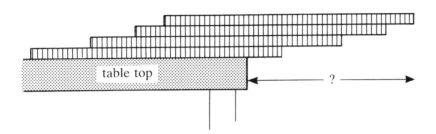

table top

?

> **Problem: how much overhang can you get?**

You are not allowed to
stick the rulers together!

Try just one ruler.
Try to explain the
result.

Try two
rulers... or three
or four

Does the bending of
the rulers affect
the answer?

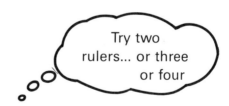

Problem:
Find the overhang for any
number of rulers

Theories?
Variables, graphs, patterns
formulae?

Centres of mass

Take a triangular 'lamina' cut out of strong cardboard. Find its centre of mass by gradually drawing three fingers together to a point of support.

What happens it you add a lump of plasticine to one corner?

Mark the point!

What if you cut out a corner, like this?

Suppose you add 10 g masses to two of the corners?

Semicircles? Quadrants?

Extensions

Check the centre of mass by hanging the lamina from various points:

Can you make a plane lamina whose centre of mass is *outside* the body?

Toppling?

Predict when your lamina will topple over the table edge.

The cable reel

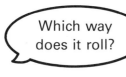

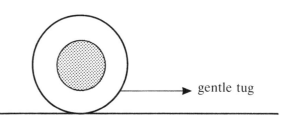

gentle tug

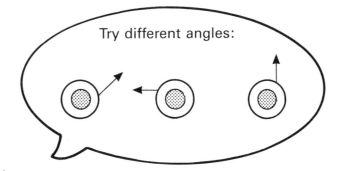

Try different angles:

■ What if . . . . .

I move it from the table onto the carpet?

The table might be sloping

■ Ideas . . . . .

I could try this with a yoyo or a cotton reel.

Why does it say 'gently'?

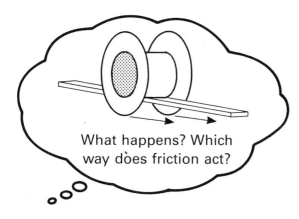

What happens? Which way does friction act?

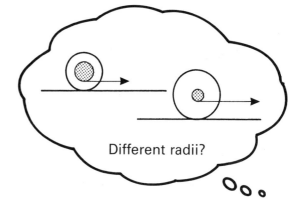

Different radii?

From Mechanics in Action

The toppling tube

Hold the roll on the table. Put one ball (about $\frac{2}{3}$ of the diameter of the roll) inside the roll, and then put another on top of the first. Let go.

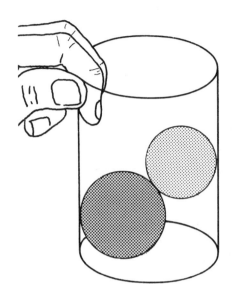

Why does
it fall over ...
... sometimes?

What pushes
it over?

Which way does
it fall

Do heavier
balls make a
difference?

Try different
rolls, heights, balls.

Problem
When will the
tube topple?

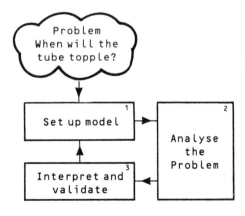

Set up model ¹	Analyse the Problem ²
Interpret and validate ³	

Try a third ball.

The ruler problem

Rest a metre ruler on your fingers and gradually begin to draw your fingers together . . . what happens?

Which finger slides first?

Set up model ¹	→	Analyse the Problem ²
↕		
Interpret and validate ³	←	

Does the ruler move?

Is the ruler in 'equilibrium'? What forces act?

What if one finger is made stickier or smoother? (Replace by a rubber or pencil)

What if you use a broom?

Or a weighted ruler?

When your finger begins to slide . . . What happens next?
- to the ruler?
- to the other finger?
- to the sliding finger?

What if the ruler is not horizontal?

The 'law' of friction

The 'law of friction states that when a body has a surface in contact with another body, a tangential friction force may act to prevent sliding of the surfaces. This friction force, F, has a maximum value, F_L, the limiting friction, which depends on the normal contact force, N. In fact, $F \leqslant F_L = \mu N$, where μ is called the coefficient of friction.

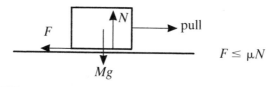

You are to validate the law of friction in an experiment.

Investigate the equilibrium of the 'block/string/mass system'.

■ How are friction and normal reaction related to M and m?

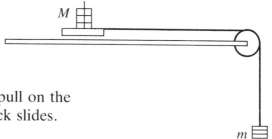

■ Masses are added to m until, eventually, the pull on the block overcomes limiting friction and the block slides.

■ Now repeat with M increased by 100 g, 150 g, 200 g . . . . .

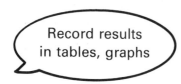

Record results in tables, graphs

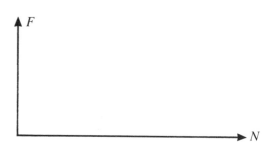

■ What can you say about the law of friction and the coefficient of friction?

Find a formula to fit your data

Angle of friction

Gradually increase the inclination of the plane until the
body slides – when $\alpha = \alpha_C$ say.

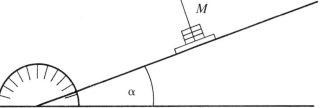

> **Investigate how α_C varies with the mass M**

■ What is the connection between the angle α_c and the
coefficient of friction μ?

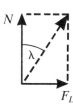

Compare
theoretical and
practical results!

■ What is the connection between the angle α and the
angle of friction λ? (The angle of friction is defined
to be the greatest angle that the resultant of F and N
can make with the normal to the contact surface. This
happens when friction is limiting, $F = F_L$.)

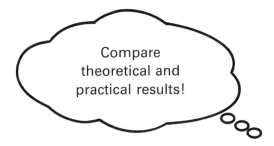

Extension

■ Sometimes bodies slide, and sometimes they
topple: investigate.

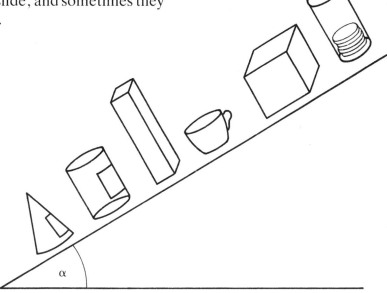

■ For a given angle of incline, say $\theta = 60°$, what can you say about *m* if the block is to remain in equilibrium?

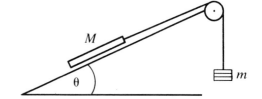

■ You should be able to find a range of values for *m* for any given *M*.

What if the angle was only 20°?

How will these values change if you double *M*?

Don't take hundreds of measurements – just describe qualitatively what happens.

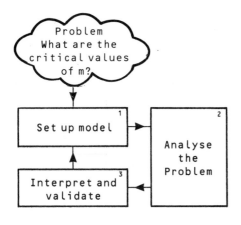

Identify interesting problems to solve.

Least force problems (2)

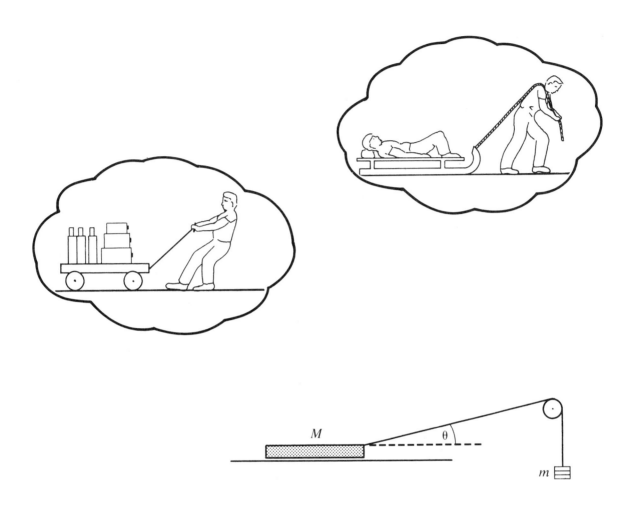

■ What is the least value of *m* required to make
the block move, for a given θ?

■ How does this least value depend on θ?

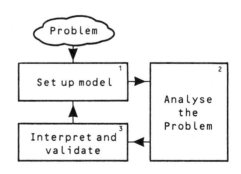

■ Is there a 'best' pulling angle for a sleigh,
sledge or trolley?

The ladder problem

> **Why is it safer for a window cleaner to have a partner to stand on the bottom rung?**

Investigate this problem with a metre rule, some plasticine and weights.

■ At what angle does the ruler slide when the weight is . . . . .

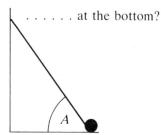
. . . . . . at the bottom?

A

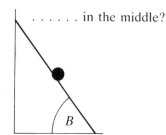
. . . . . . in the middle?

B

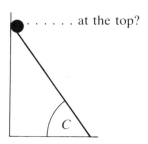
. . . . . . at the top?

C

■ Is there much difference between *A*, *B* and *C*?

What if the floor is carpetted rather than lino or wood?

What if the wall is carpetted rather than smooth?

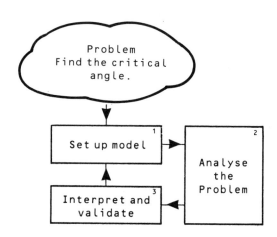

The dangerous sports club problem

You will have seen film of members of a 'Dangerous Sports Club' performing a stunt where, having tied themselves with elastic rope to a bridge, they jump off.

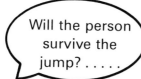

Will the person survive the jump? . . . . .

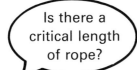

Is there a critical length of rope?

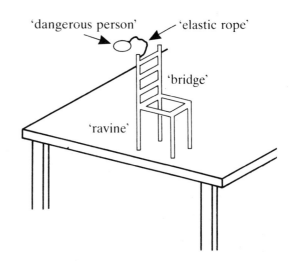

Investigate this with the aid of elastic (e.g. shirring elastic), masses (e.g. fishing weights, 10 g or 100 g masses), a metre ruler, some tables and chairs.

Does the mass of the person matter?

How 'elastic' is the elastic rope?

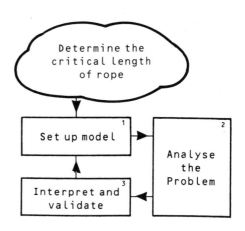

Distance, speed and acceleration (1)

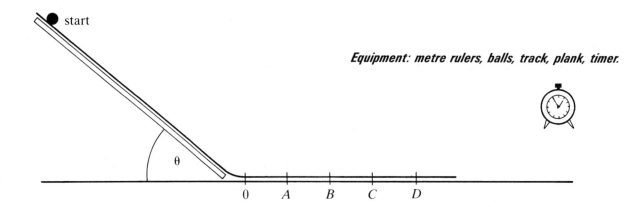

start

Equipment: metre rulers, balls, track, plank, timer.

0   A   B   C   D

Problem: Measure the speed of the ball after rolling down the slope a given distance to O.

Time the balls to *O, A, B, C, D.*

How does this speed depend on the time it has rolled down the slope?

Vary the start points and calculate speeds.

Can you find the acceleration of the ball down the slope?

. . . . . graph the speed against the time.

Extension

What is the effect of varying the angle of slope θ?

Distance, speed and acceleration (2)

Equipment: metre rulers, rubber balls, snooker balls.

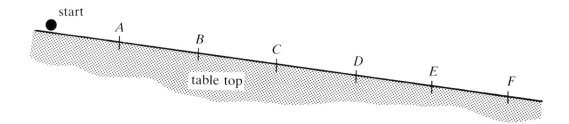

plan view

rulers
as
guides

> **Problem: to describe how speed increases as the ball rolls down the slope.**

We could plot speed against time, or against distance . . . . . What does this tell us?

We can measure distance and time then deduce speed.

How will the graphs be affected by . . . . .
. . . . . a change of slope?
. . . . . a more massive ball?
. . . . . a rougher surface?

Could we take more accurate data using a video and electronic timer?

C.B. de Mille

Connected masses problem

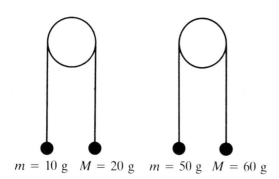

$m = 10$ g $M = 20$ g $m = 50$ g $M = 60$ g

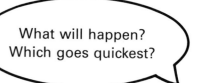

■ Release the two systems simultaneously

> What will happen?
> Which goes quickest?

> **Investigate the relationship between the two masses and the resulting acceleration.**

> How do you explain this?

> Draw force diagrams, apply Newton's laws and calculate acceleration . . . . .

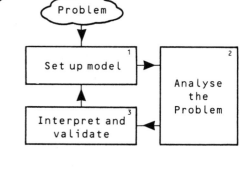

■ Does the theory explain reality?

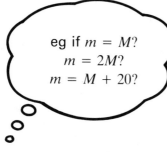

> eg if $m = M$?
> $m = 2M$?
> $m = M + 20$?

> What if . . . . .
> . . . . . $m = 0$, $M \neq 0$?
> . . . . . $m = M = 0$?

Extensions

■ What assumptions did you make in your mathematics? Test their 'accuracy'.

■ Try some other 'connections'.

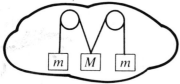

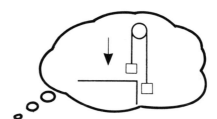

The high road and the low road

Equipment: 'Streak' car racing kit, various balls (at least two the same).

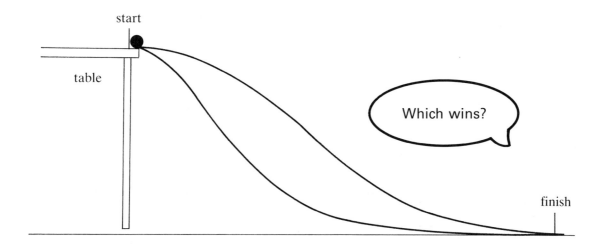

> **If I release the two balls down the two tracks simultaneously from the same height, which one gets to the floor first? Explain, using speed–time graphs.**

Extension

■ Try to model the shape of the track in various ways and calculate the times of descent: eg

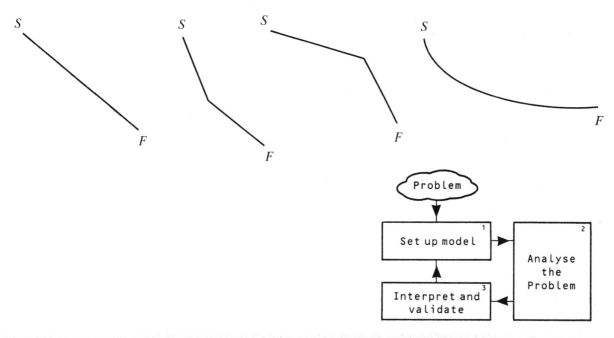

Looping the loop (1)

Equipment: 2 'Streak car racing kits', cars, marbles, balls.

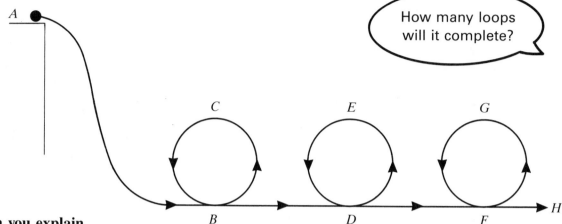

How many loops
will it complete?

■ **Can you explain**

Why it needs more
initial height to do
two loops than it does
to do one loop?

What happens
to its energy?

Why the ball
moves fastest/slowest
at *B, D, F / C, E, G*?

The difference
between a car and
a ball?

Why the
track moves?

> **Predict, for a given initial height, how many loops
> a ball will complete. Sketch a graph of the ball's
> speed throughout its motion.**

What is the speed of the ball
at the top of the first loop?

What is the speed of the ball
at the bottom of the first loop?

What is the difference
between the two?

Extension

■ Using an appropriate estimate of energy lost per loop, repeat the
above estimates and calculations for a ball which just completes
two loops.

A projectile problem (1)

Simulate a projectile by rolling a ball
on an inclined table. Stick sugar
paper on the table.

When you have tried a few 'dry runs',
dampen the ball and you will get a
trace of its path: mark the path with
a felt tip pen.

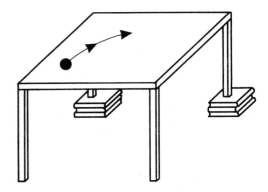

> **Problem: find a formula which describes the path
> of the ball across the table.**

Cut along
the path and
get an 'instant
graph'.

What sort
of formula gives
this shape?

Introduce
variables.

Collect
data.

Try some formulae;
a graphic calculator
may help.

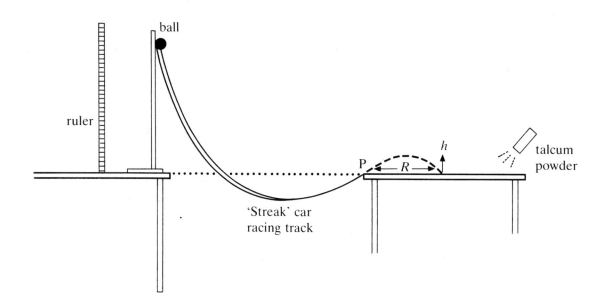

Control the speed of projection by varying the initial height of the
ball above the projection point, *P.* Control the angle of projection by
varying the shape of the track.

■ Validate the theory of projectiles . . . .

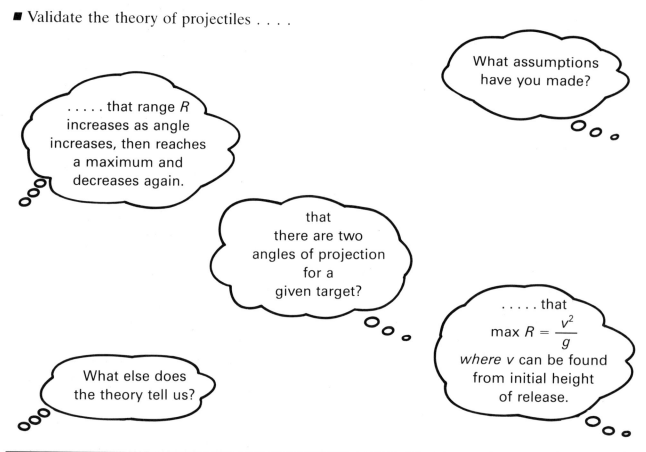

A projectile problem (3)

Equipment: Ball / coin / ball bearing. A means of projecting with constant speed (e.g. spring-loaded toy, coin thrower, etc).

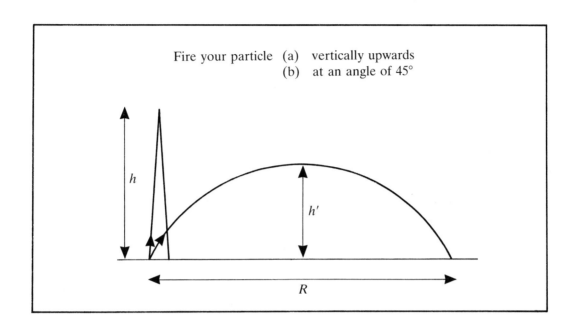

Fire your particle (a)   vertically upwards
                         (b)   at an angle of 45°

Verify that R is approximately $2h$. Why is this?

Calculate the velocity of projection

Can you get a bigger range than R?

Make further predictions and test them.

Predict the range if fired at an angle of 30° or 60°.

Pennies on a turntable

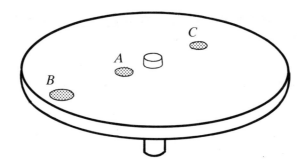

Set the turntable rotating at a steady speed.

What is this angular speed .....
..... in rpm?
..... in radians per second?

Which penny moves fastest
and which slowest?
Calculate their speeds in
metres per second.

Place two pennies so that
one has exactly double
the speed of the other.

What happens to the
speeds if you change
the rotation speed?

In which direction
is *B* moving at a
given instant?

Extension

True or false?

- Each penny has constant speed.
- Each penny has constant velocity.
- The velocity of a penny is always changing.
- The displacement of a penny from the axis of rotation is constant.
- The angular acceleration is zero.
- The actual acceleration is constant.
- There is a resultant force on each penny causing acceleration.

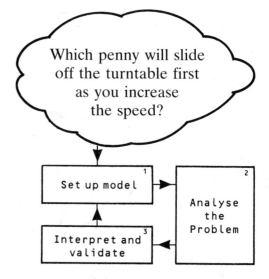

Which penny will slide
off the turntable first
as you increase
the speed?

Conical Pendulum (1)

Set the bob moving in a horizontal circle – like this:

What do you have to do to keep it going in a circle? How do you make it go faster?

What **force** does your hand feel? Why?

What forces act on the **bob**? . . . and the **string**?

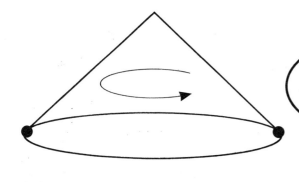

What do you have to do to get **small** circles? . . . . . **large** circles?

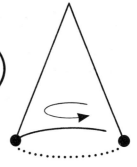

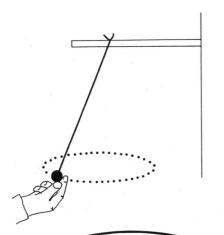

Can you make it go in circles by projecting it? Why is it so difficult?

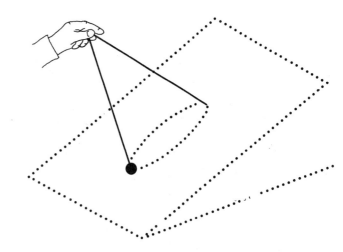

Can you make it go in circles in an inclined plane? Why not?

Banking

Attach a banked section
of track to a turntable.

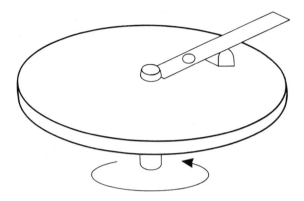

**Investigate the effect of the banking angle
on the tendency to slide.**

What effect does
banking have on the
critical angular speed?

Will the penny **always**
slide outwards for some
critical speed?

Are there any
other critical speeds?

For a given banking
angle, is there an ideal
angular speed for which
friction is zero?

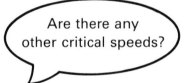

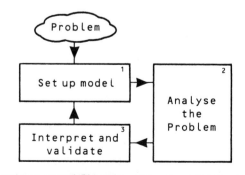

From *Mechanics in Action*

The rotor

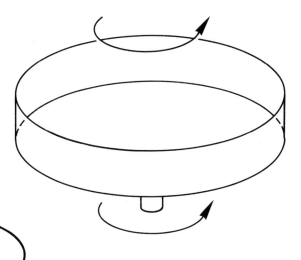

What happens when the floor is removed?

Is there a critical speed? What is it?

... and what does it depend on?

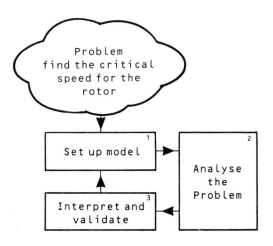

Problem
find the critical speed for the rotor

Set up model ¹

Analyse the Problem ²

Interpret and validate ³

You can explore this practically with a penny on a track on a turntable like this.

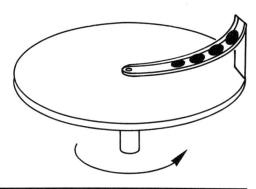

Conical pendulum (2)

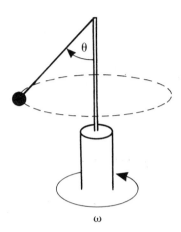

Investigate the effect of increasing the angular speed
from zero to top speed.

What happens
to the angle θ
as ω increases?

What happens to
the height of the
bob's path?

Is the mass
of the bob significant?

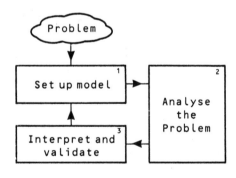

Can you get the bob to 'lasso'?

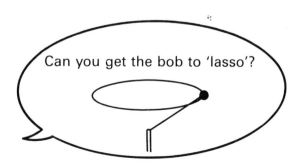

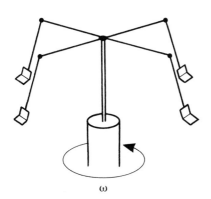

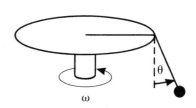

Is the mass in
the chair significant?

What happens to
the angle θ as speed
ω increases?

Set up model ¹	**Analyse the Problem** ²
Interpret and validate ³	

What about the
length of the
string?

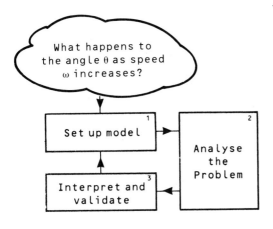

Do inner and
outer chairs swing
out at the same angle?

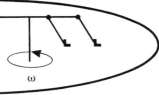

Looping the loop (2)

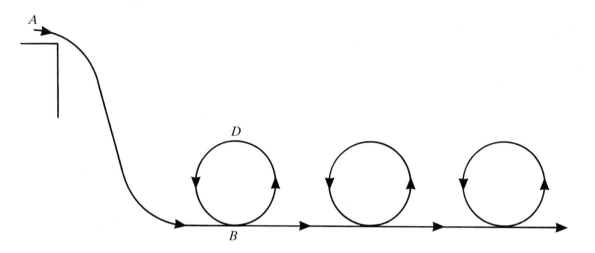

Try releasing cars, marbles, balls down the track.

What problems occur to you that you might solve using a mechanical model?

What is the least height of release for a ball to . . . . .

What will happen to the particle once it is released?

Problem

Set up model ¹

Analyse the Problem ²

Interpret and validate ³

At which points of the loop can the car fall off?

Is this track smooth?

Does the shape of approach track or size of loop matter?

Is the body a particle?

Wind-up

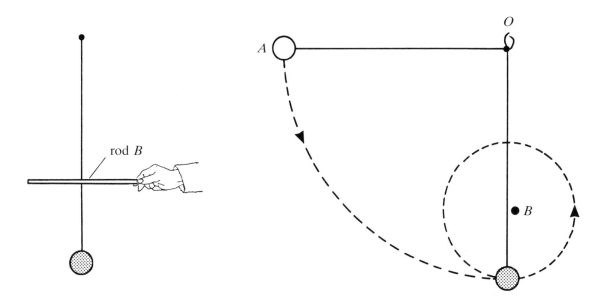

A weight on the end of the string is released from
the horizontal *A*, by one student and falls until the
string hits the rod *B* held by another student. Will
the weight complete a vertical circle about *B*?

It depends on
the length of
the string.

It depends where
I hold the rod *B*.

Does it depend
on the radius of
the circle about
B?

Shouldn't it
keep rotating
about *B* for
ever?

. . . . . it depends
on the thickness
of the rod *B*?

Problem

Set up model ¹

Analyse
the
Problem ²

Interpret and ³
validate

From *Mechanics in Action*

worksheet
A stop–go phenomenon

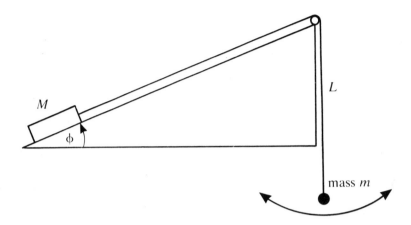

Attach a mass *m* such that the block is just about to slide up the plane, but does not quite slide. Set the mass oscillating like a simple pendulum. You should find (for some values of *m*, *M*, φ, amplitude and string length) that the block lurches up the plane.

At which point does the block *M* begin to slide . . . stop sliding?

Why does the length and amplitude of the pendulum affect things?

■ Can you find a condition on *m*, *M*, *l*, φ, and amplitude which will allow this phenomenon to take place?

■ What would happen if the plane were smooth?

■ What would happen if the block were just about to slide **down** the plane?

From Mechanics in Action

Cake tin

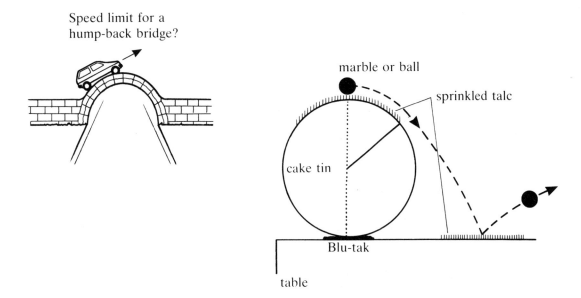

A marble is slightly disturbed from the top of the cake tin.

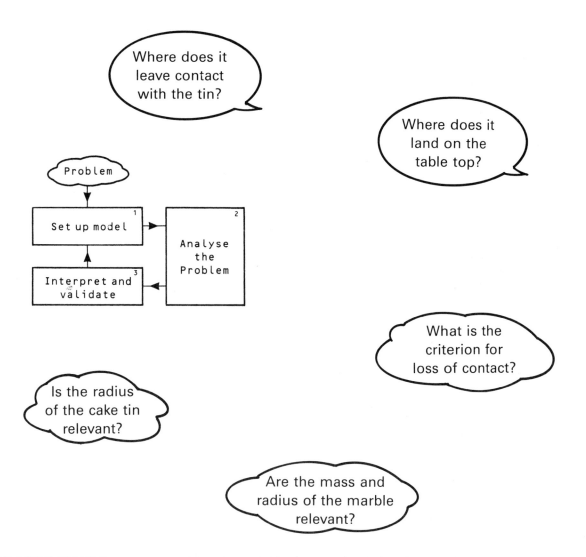

Timing oscillations

Equipment: a stopwatch; a retort stand; some springs (identical); masses

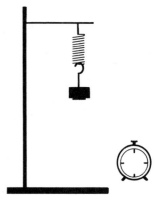

Investigate the rate of oscillation or time period of the oscillator.

> How do you measure 'rate of oscillation'?
> How do you measure the time period?

> What does it depend on?
> What can I vary?

■ Sketch a graph showing how the time period of the mass changes with mass

> Control variables.
> Minimise errors.
> Tabulate and graph results.
> Find rules and formulae.

Extensions

> How many different ways can I find of getting a one second time period?

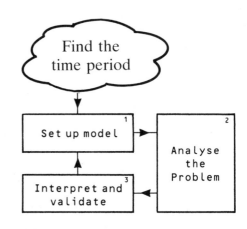

> Find the time period

```
Set up model  ──►  Analyse
   1               2  the
                      Problem

Interpret and  ◄──
validate
   3
```

Modelling a mass spring oscillator

Set up a mass spring system with a large extension. The idea
is that it has a long time period, and moves slowly through
a large distance from *A* to *B* and back again.

ceiling

> **Problem: to find a function which describes the
> displacement of the mass (from the ground, say) as
> a function of time.**

Start with a displacement–time sketch-
graph over about a minute.

h

height of *B*

height of *A*

t

How long does an oscillation take?
Does this 'time period' stay
constant?

When and where is it
moving fastest? . . . . .
slowest: . . . upwards?
. . . . . downwards?

When is the mass
in equilibrium?

When is the tension
greatest? . . . . . least?

When is the acceleration
upwards? . . . . . downwards?

What is the
resultant force
at any point?

Find a function *h*(*t*) which fits your graph.
Does it have the right amplitude? . . . . . time
period? . . . . . turning points?
Is the same true of *dh*/*dt* and the second
derivative?

B

X

A

Hooke's law

Hooke's law states that the 'tension in the spring' is proportional to its extension

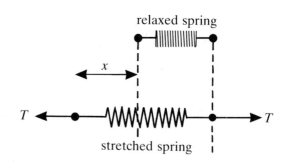

relaxed spring

stretched spring

$$T \propto x \text{ or } T = kx.$$

The constant of proportionality, k, is known as the spring stiffness.

Investigate this law using mass, springs, ruler and retort stand.

$m = 0$ implies $x = 0$?

If you double m . . . . . ?

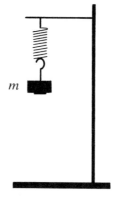

m

Is the graph of T against x a straight line?

What m would be required to stretch the spring through, say, 100 metres?

Either: Hooke's law is valid (within reason, or within limits). In this case, calculate the stiffness of the spring in newtons per metre.

or: you must propose an improved model of tension in a spring, such as $T = kx^3$ or $T = kx + c$.

Stiffness and elasticity

Equipment: 4 or 5 identical springs, ruler, masses and stand;
various lengths of elastic.

1 Compare the 'stiffness', k, of 1, 2, 3, . . . or
more, identical springs connected in series.

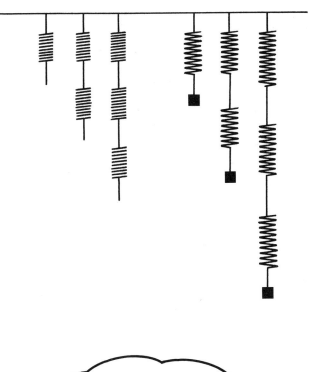

'Identical'
springs?

Suppose
they are connected
in parallel?

Is there a connection
between k and length?

2 How does the stiffness of a length of elastic
depend on its natural length?

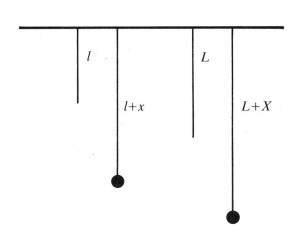

Can I assume
that the elastic is
'uniform'?

Does stretching
the elastic change
its elasticity?

Amplitude decay

Take a mass spring oscillator and attach a damper.
What do you notice
- about the amplitude of the oscillations?
- about the time period of the oscillations?

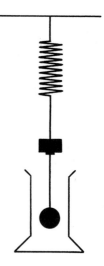

What does the sketch graph of the displacement look like now?

What can you say about the forces on the damper? (Try pulling the damper up and down in liquid.)

How does the damping effect depend on
. . . . the mass on the oscillator?
. . . . . the size of the damper?
. . . . . the liquid in which the damper moves?
. . . . . ?

Extension

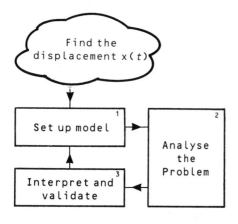

Find the displacement x(t)

Set up model ¹ → Analyse the Problem ²

Interpret and validate ³

Is it possible to set up the system so that there are no oscillations at all?

Resonance

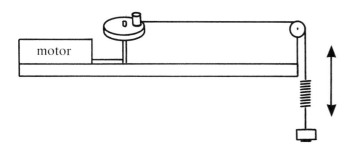

Set up a forced oscillator, using an offset cam and a mass spring oscillator.

Vary the rotational speed of the cam. At very high or very low revolutions the oscillator seems not to notice!

Find the resonant frequency.
How is the resonant frequency affected by the mass on the oscillator?

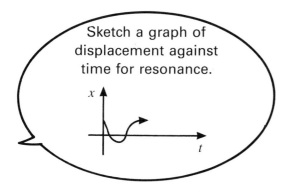

Sketch a graph of displacement against time for resonance.

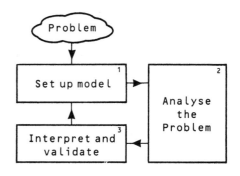

Extensions

■ What happens **near** the resonant frequency?
■ Can you find a 'beats' motion?

Explain resonance in terms of 'energy' in the system. Why does the work done by the 'forcing' motor have little or no effect except at the resonant frequency?

Explain 'beats' in terms of the work done by the motor on the oscillator.

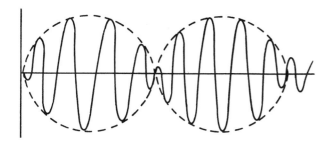

Simple pendulum

What does the theory say about the time period of simple pendulum of length l and mass m?

The theory of simple harmonic motion says: 'the time period T of a simple pendulum is

$$T = 2\pi \sqrt{\frac{l}{8}}$$

where l is the length of the pendulum'.

> **Interpret and validate this statement.**

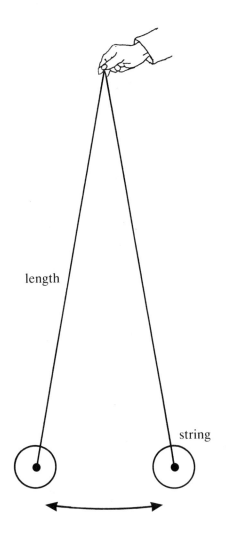

length

string

> If I double the pendulum length, then the time period will . . . . . ? Check this!

> How does the time period change with mass? Check this!

> How can we calculate g? Do it! Does this agree with the text books?

> How does the time period change with the amplitude of the oscillation? Check this!

> What would happen . . . . on the moon? . . . in orbit?

A compound pendulum

Equipment: Metre rule with holes drilled at 5 cm intervals; stopwatch.

> **Problem: how is the time period of the oscillations affected by the placing of the needle?**

needle

Tabulate and graph some results.

Is there a maximuim or minimum time period?

Problem

| Set up model | ¹ | → | Analyse the Problem | ² |

↑

| Interpret and validate | ³ | ← |

Extensions

1 Attach masses to the ruler.

2 Try to obtain a given time period (eg 1 second) by attaching weights (eg coins) to the appropriate parts of the ruler.

Bouncing ball (1)

**Investigate the relationship between the rebound
height and the drop height for a ball.**

How accurate and
consistent are
your measurements?

Can you find a
rule or formula?

Make more
predictions and
test them.

Predict the height
after 2, 3 or 4
bounces.

drop

rebound

The law of restitution says that when two bodies moving
in a straight line collide, their relative speed after impact
is proportional to their relative speed before impact.
The constant of proportionality, e, is called the
coefficient of restitution, and $0 \leqslant e \leqslant 1$. Thus

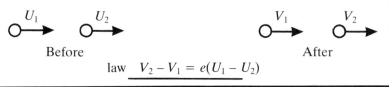

$$U_1 \qquad U_2 \qquad\qquad\qquad V_1 \qquad V_2$$

Before   After

law $V_2 - V_1 = e(U_1 - U_2)$

■ Validate the 'law of restitution' for impact in this case,
and calculate e for your ball.

What is the connection
between impact speed
and drop height?

How valid
is the law?

Bouncing ball (2)

A ball is released with some sideways motion and bounces a number
of times . . . . .

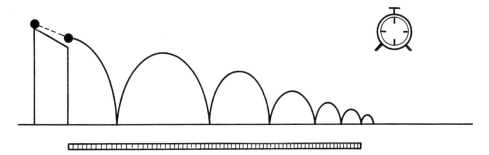

Investigate one of the following
- the distances between bounces,
- the heights of the bounces.
- the time it takes to stop bouncing.

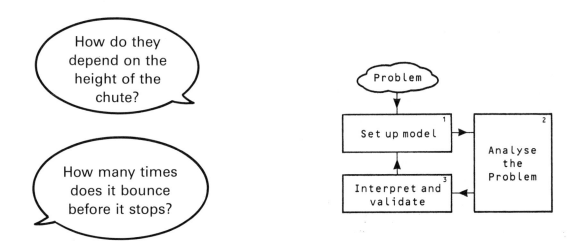

How do they
depend on the
height of the
chute?

How many times
does it bounce
before it stops?

Problem

Set up model	¹
Analyse the Problem	²
Interpret and validate	³

Extensions

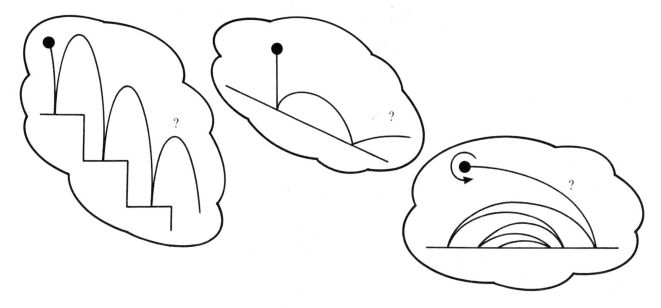

Newton's cradle

In each case you should guess first, test your prediction in practice, then explain your result mathematically in terms of momentum and energy.

before . . . . .
after?

before . . . . .
after?

before . . . . .
after?

before . . . . .

Any more ideas?

What if . . . . .

- you stick two balls together?
- you add some lead to one of the balls?
- you put the cradle on a slope?
- you leave the balls to impact several times?
- the balls were so sticky that they stuck together at impact?

Rebound

'When a snooker ball hits
the cushion, it rebounds at
the same angle just like
reflected light.' Is this true?

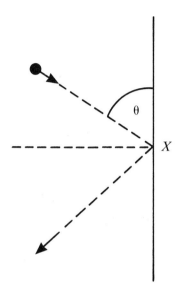

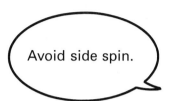

Avoid side spin.

Experiment

■ Roll a ball along the floor to hit the wall at
a marked point, X, so that you can chalk
the incident and reflected line.
■ Measure appropriate angles, and repeat for
various θ.
■ Make tables, graphs and formulae.
■ Produce a theory involving factors such as
mass, coefficient of restitution, speed of
impact, etc.

. . . . . even
better on a snooker
table

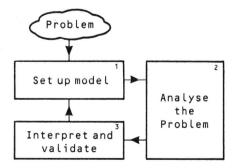

Extension

■ What if the ball is spinning?

The superball as a deadly weapon

You need: superballs of different sizes; various other balls, e.g. tennis and soccer balls.

Drop the balls from about 1 metre:

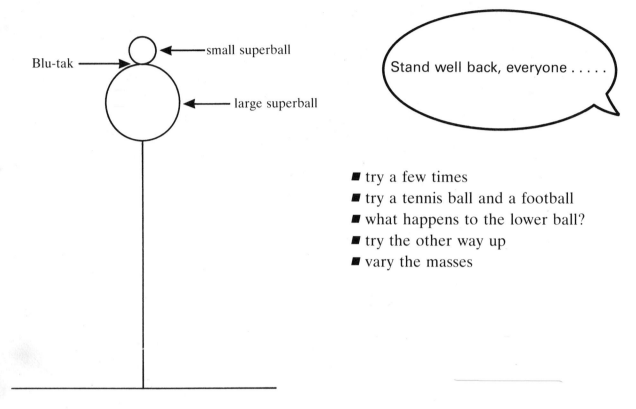

Stand well back, everyone . . . . .

■ try a few times
■ try a tennis ball and a football
■ what happens to the lower ball?
■ try the other way up
■ vary the masses

■ How can this be explained? What assumptions do you need to make to apply principles to the problem? Remember: keep things *simple* at first.

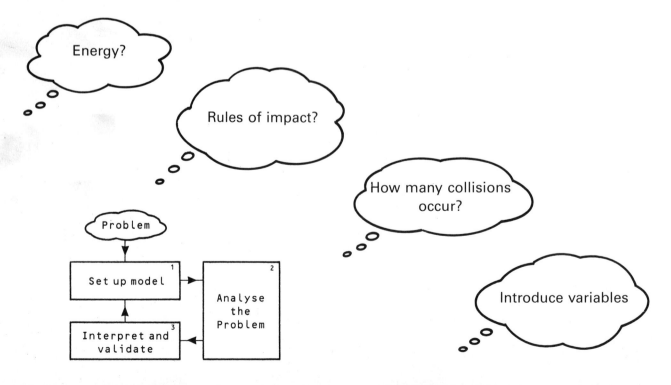

Energy?

Rules of impact?

How many collisions occur?

Introduce variables

Problem

Set up model ¹

Analyse the Problem ²

Interpret and validate ³